ALFRED RUSSEL WALLACE

Alfred R Wallace

ALFRED RUSSEL WALLACE

A REDISCOVERED LIFE

MICHAEL A. FLANNERY

SEATTLE   DISCOVERY INSTITUTE PRESS   2011

Description

Alfred Russel Wallace: A Rediscovered Life is a new biography of the co-discoverer of the theory of evolution by natural selection and one of the nineteenth century's most intriguing scientists. Its provocative thesis is that Wallace, in developing his unique brand of evolution, presaged modern intelligent design theory. Wallace's devotion to discovering the truths of nature brought him through a lifetime of research to see genuine design in the natural world. This was Wallace's ultimate heresy, a heresy that exposed the metaphysical underpinnings of the emerging Darwinian paradigm. Biographer Michael A. Flannery is Professor and Associate Director for Historical Collections at the Lister Hill Library of the Health Sciences, University of Alabama at Birmingham (UAB) and editor of *Alfred Russel Wallace's Theory of Intelligent Evolution* (2008).

Publisher's Note

This book is part of a series published by the Center for Science & Culture at Discovery Institute in Seattle. Previous books include *The Deniable Darwin* by David Berlinski, *In the Beginning and Other Essays on Intelligent Design* by Granvile Sewell, *God and Evolution: Protestants, Catholics, and Jews Explore Darwin's Challenge to Faith*, edited by Jay Richards, and *Darwin's Conservatives: The Misguided Quest* by John G. West.

Library Cataloging Data

Alfred Russel Wallace: A Rediscovered Life by Michael A. Flannery (1953–)

166 pages, 6 x 9 x 0.38 inches & 0.56 lb, 229 x 152 x 9.8 mm. & 0.254 kg

Library of Congress Control Number: 2010941745

BISAC: BIO015000 Biography & Autobiography/Science & Technology

BISAC: SCI027000 Science/Life Sciences/Evolution

BISAC: SCI034000 Science/History

ISBN-13: 978-0-9790141-9-2   ISBN-10: 0-9790141-9-0 (paperback)

Publisher Information

Discovery Institute Press, 208 Columbia Street, Seattle, WA 98101

Internet: http://www. discovery.org/

Published in the United States of America on acid-free paper.

First Edition, First Printing. January 2011.

Praise for *Alfred Russel Wallace: A Rediscovered Life*

FLANNERY'S RIVETING TALE OF REDISCOVERY PROVIDES CONVINCING new evidence that Alfred Russel Wallace—the acknowledged co-discoverer of evolutionary theory—supported an argument from design for all forms of life which, in many ways, anticipated modern intelligent design thinking. This fascinating work of intellectual history recasts a new, more complete and lasting image of the once all too elusive Wallace.

Philip K. Wilson, MA, Ph.D., Historian of Medicine and Science
Professor of Humanities and Science, Technology & Society
Director, The Doctors Kienle Center for Humanistic Medicine
Penn State College of Medicine

MICHAEL FLANNERY HAS WRITTEN A SUPERB BOOK THAT IN ITS PASSION and subjective honesty offers a cogent and articulate defense of Alfred Wallace's theory of intelligent design, with all its moral and ethical implications, as a counterpoint to the materialistic worldview that came to be known as Darwinism.

John S. Haller, Ph.D.
Emeritus Prof. of History and Medical Humanities
Southern Illinois University, Carbondale

THIS BIOGRAPHY OF ALFRED WALLACE BY MICHAEL FLANNERY IS THE most important new book I have read in years. The immense attention focused on Charles Darwin by evolution historians has unfortunately overshadowed Wallace, whose life was arguably more fascinating and insightful. Unfortunately views that are offered on Wallace today are often from Darwinist perspectives. Flannery remedies this imbalance with his story of Alfred Wallace that brings an entirely new light to the theory of evolution. In this corrective against the familiar but erroneous casting of Wallace as a miniature Darwin, Flannery artfully brings out the stark contrast—even down to their final works—between the evolution co-founders. But this volume is not merely a look back. Like any good history, Flannery's tells us where we are, and how we got here.

From their early years onward, Wallace and Darwin existed in different worlds. Their paths intersected at evolution, but they approached and departed that intersection with many different perspectives. Flannery provides a broader context than is usually found in such histories and in convincing detail demonstrates the influences and connections to today's discussion. Neither idolizing Wallace nor minimizing Darwin, Flannery provides a much needed balanced view that leaves us with a richer understanding of our ideas on origins.

Cornelius Hunter, Ph.D.
Author of the books Darwin's God, Darwin's Proof, and Science's Blind Spot

IN THIS LUCIDLY WRITTEN BOOK MICHAEL FLANNERY SHOWS THAT Alfred Russel Wallace, the co-discoverer of the theory of evolution by natural selection, thought the theory was incomplete—the guidance of a higher power was needed to explain nature. Wallace's ideas show strongly that the godless view of evolution taken by so many modern evolutionists is not forced on them by the evidence; rather they assume it in spite of the evidence.

Michael Behe, Ph.D., Professor of Biochemistry, Lehigh University
Author, Darwin's Black Box and The Edge of Evolution

FOR TOO LONG, ALFRED WALLACE'S CONTRIBUTIONS TO THE SCIENCE of evolution have been ignored or grossly mischaracterized. Why? Largely because he provided both a coherent criticism of Darwin and Darwinism and a theist-friendly alternative account of evolution. The secular intelligentsia sided with Darwin because Darwinism provided them with a view of evolution that handily eliminated God. They succeeded in making Darwinism the default view of evolution, both in history books and biology textbooks. As a result, our understanding of the history of evolutionary theory and our understanding of evolution itself have suffered. Wallace's account of evolution, if it had received a fair hearing, would have (I believe) won the day, and our understanding of the history of evolutionary theory and evolution itself would be much different today. But it is not too late. Wallace may yet become the beginning point of intellectual renewal, and Michael Flannery's fine biography of Alfred Russel Wallace will go a long way in bringing that much-needed revolution about.

Benjamin Wiker, Ph.D.
Author of The Darwin Myth

HISTORIAN MICHAEL FLANNERY'S ACCOUNT OF ALFRED RUSSEL WALlace's life work is a lucid sketch of the scientific and philosophical controversies over evolution in the mid-nineteenth century... Wallace observed, to Darwin's chagrin, that man's intellect—his reason, his artistic and musical ability, his wit, his talent, and most of all man's moral sense—must be caused by an "Overruling Intelligence" that guided evolution. Wallace insisted that man's mind was created by a Mind. Flannery's book is a concise and eloquent exploration of Wallace's genius and of his rejection of Darwin's implicit materialism and atheism. These differences persist in our modern debate about origins, and today Wallace's views may well be advancing. Flannery's superb book provides the reader with indispensable insight into the earliest squalls in the modern tempest over Darwin's theory and intelligent design.

Michael Egnor, M.D., Professor and Vice-Chairman
Department of Neurological Surgery
Stony Brook University Medical Center

FLANNERY'S BOOK IS A WELCOME ADDITION TO ANY BOOKSHELF THAT offers a window into evolution. It comes down to this: Darwin was following evidence that supported a materialist theory he already espoused; Wallace was following evidence that shed light on the nature of nature. Popular culture chose Darwin, and the rest is the stale Darwin worship promoted in popular culture and academy alike, as an alternative to engagement with the facts.

Denyse O'Leary
Co-author of The Spiritual Brain

CONTENTS

ILLUSTRATIONS

PREFACE

IN A SENSE, THIS BOOK HAD A VERY LONG GENESIS. IT STARTED YEARS ago when I was an undergraduate reading books like Jacques Barzun's *Darwin, Marx, Wagner: Critique of a Heritage* (1941), Gertrude Himmelfarb's *Darwin and the Darwinian Revolution* (1962), and a very different kind of writing (a recent release at the time) William Irwin Thompson's *At the Edge of History* (1971). It is from these books and certain young and impassioned faculty at an equally young Northern Kentucky State College (now NKU) that I learned true critical thinking. Fast forward to a few years ago and an initially casual reading of Alfred Russel Wallace's *The World of Life* and the ingredients for the present study were already put in place. I thank Bill Dembski for seeing the value of turning that casual reading into serious analysis with my *Alfred Russel Wallace's Theory of Intelligent Evolution: How Wallace's World of Life Challenged Darwinism* (2008); it was truly a prolegomenon to the present biography, which brings me to those helpful to the work at hand. My deepest heartfelt thanks goes to John West for his interest in and commitment to this entire Wallace project. I am especially appreciative of his careful reading of this book's initial manuscript. His thoughtful comments and suggestions made for a much stronger work. I also want to thank everyone at the Discovery Institute for their help in making this a reality and in finally letting Wallace's story be told. Also, I am thankful for those colleagues who read this biography in manuscript and offered their assessments, many included in the preliminary pages of this work. But I have even greater appreciation for my wife Dona who tolerated with admirable equanimity her absentee husband wedded to his keyboard through many weekends. Lastly, honorable mention goes to Ebbet the cat, my silent partner through most of this book's writing; it was good to have a friend by my side during the dense solitude of historical mining.

INTRODUCTION

FOR YEARS ALFRED RUSSEL WALLACE WAS LITTLE MORE THAN AN obscure adjunct to Charles Darwin's theory of evolution. Remembered only for prompting Darwin to write *On the Origin of Species* in 1859 by sending Darwin his own letter proposing a theory of natural selection, Wallace was rightly dubbed by one biographer "the forgotten naturalist."[1] In 1998 Sahotra Sarkar bemoaned Wallace's "lapse into obscurity," noting, "At least in the 19th century literature, the theory of evolution was usually referred to as 'the Darwin and Wallace theory'. In the 20th century, the theory of evolution has become virtually synonymous with Darwinism or neo-Darwinism."[2] While the complaint still has a ring of truth, a decade of recent interest in Wallace has done much to bring him back from history's crypt of forgotten figures. This shouldn't suggest unanimity of opinion, however. Some regard him as a heretic, others as merely a misguided scientist-turned-spiritualist, still others as a prescient figure anticipating the modern Gaia hypothesis.[3] Perhaps Martin Fichman's phrase hits closest and most persistently to the truth—"*the elusive Victorian.*"

Can the *real* Wallace be found? If so, what might we learn in that rediscovery? It is worth stating the thesis here at the outset: Wallace, in developing his unique brand of evolution, presaged modern intelligent design theory. Certainly no Christian creationist, Wallace's devotion to discovering the truths of nature brought him through a lifetime of research to see genuine design in the natural world. And this indeed became Wallace's heresy, a heresy that exposes the metaphysical underpinnings of the triumphant Darwinian paradigm more than it does Wallace's commitments to spiritualism or science. The image of Darwinism reflected in the image of natural selection's co-discoverer is indeed an interesting one. But it all began oddly enough in an obscure village far from the seats of learning or science.

MY BIRTHPLACE. KENSINGTON COTTAGE, USK

FREE LIBRARY, NEATH
(Designed by A. R. Wallace. 1847)

1.

THE FIRST TWENTY-FIVE YEARS

EARLY LIFE FROM USK TO NEATH, 1823–1848

ALFRED RUSSEL WALLACE WAS BORN NEAR THE WELSH/ENGLISH border town of Usk on January 8, 1823, the eighth addition to the family of Thomas Vere and Mary Ann Wallace. Although Thomas had been trained in the law, there is no indication of his actually practicing. While his father had descended from the Scottish rebel Sir William Wallace and his mother from French Huguenots, they regarded themselves as devoted English subjects and devout Anglicans. Alfred's family circumstances have sometimes been described as impoverished, but struggling and declining middle class probably best describes the family into which he was born. As a bachelor Thomas Vere Wallace decided to forgo the life of an attorney, preferring to use his £500 annual inheritance enjoying "himself in London and the country, living at the best inns or boardinghouses, and taking part in amusements of the period, as a fairly well-to-do, middle-class gentleman."[4] But such experiences were not conducive to maintaining married life with a sound financial footing, and within a decade of his marriage to Mary Ann his substantial means had been wrecked upon "the most risky of literary speculations."[5] Alfred Wallace's background was similar to that of Thomas Henry Huxley. The fathers of both had trained in the professions and both had slipped off the social ladder, but as historian Iain McCalman has noted, "Thomas Wallace had fallen harder and further than George Huxley."[6]

The family's inexorable slide was halted when Mary Ann received a modest inheritance with the death of her stepmother in 1828, whereupon the Wallaces moved to Hertford where they continued to reside until around 1837. It was at Hertford that Wallace received his only formal education at a grammar school run by a headmaster described by his student as "a rather irascible little man named Clement Henry Crutwell."[7] Alfred's instruction in Latin grammar and the other standard classics of the day, lessons instilled with judicious canings, knuckle-raps, and ear-boxings, concluded at the end of 1836.

Renewed financial constraints forced Alfred's father to remove him from school and send him off to London to live with his brother John who was apprenticing to a master builder name Mr. Webster. It is here that Wallace's biographer Ross A. Slotten finds the roots of his subject's social radicalism and for good reason.[8] In the summer of 1837 Alfred would move to Bedfordshire where his older brother William would teach him the life and skills of a land-surveyor, but not before the young teenager drank from the intellectual wells of radical socialist Robert Owen. It was at evening meetings of the "Hall of Science" that Alfred learned from the works of Robert Dale Owen (eldest son of the radical firebrand) that "the orthodox religion of the day was degrading and hideous, and that the only true and wholly beneficial religion was that which inculcated the service of humanity, and whose only dogma was the brotherhood of man. Thus was laid," Wallace recalled years later, "the foundation of my religious scepticism."[9]

Alfred Russel Wallace via Owenite socialism became duly radicalized. But Bedfordshire would begin a more significant education for the youth; the very process of learning the surveyor's trade forged a lasting bond with the land. It was during these next seven years that he would discover his love of solitude amidst nature, forge an intimate bond with the land that only a surveyor can possess, and take Sunday tours of the countryside with an inexpensive botanical field manual at his side. Wallace began his science studies in earnest with the land as his lecture hall. The surveyor's life is not a stationary one and the Bedfordshire stay was not long. In autumn of 1839 Alfred left with his brother for Wales again, this time Radnorshire. The opportunity to survey a parish in Glamorganshire in autumn of 1841 would bring the two enterprising surveyors to the Neath Valley.

Upon turning twenty-one an opportunity presented itself. The Reverend Abraham Hill, headmaster of a school in Leicester, offered the young man a teaching position. Here Wallace found a substantial library where he read of von Humboldt's travels in South America, Prescott's histories of Mexico and Peru, and, most importantly, Malthus's *Essay on the Principle of Population*, the book he admitted "twenty years later gave me the long-sought clue to the effective agent in the evolution of organic species."[10] Similarly, it was the same reading of Malthus that prompted Darwin to proclaim, "I had at last got a theory by which to work."[11] It was during this time that Wallace met Henry Walter Bates, "an enthusiastic entomologist," whose interest in beetles and butterflies would arouse in the weekend botanist an interest in nature's animal life. It was also here that Wallace was introduced to mesmerism and phrenology, both in their day considered cutting-edge (if rather controversial) science. The intellectual ferment he enjoyed in Leicester was cut unexpectedly short when his brother William died suddenly (probably of pneumonia). Wallace's opportunities for learning in a community charged with and by new ideas, his meeting with Bates, and his access to a good library (a library that introduced him to Malthusian theory), all brought him to the inescapable conclusion that his two years in Leicester were the most important of his early life. But his brother's death at Neath forced his return to Wales early.

There is a tendency to downplay Wallace's youthful experiences in Wales; Wallace certainly did so in his autobiography *My Life* (1905) written sixty years later. But R. Elwyn Hughes has uncovered important details of those years, forcing a reassessment.[12] Hughes notes that Wallace's first known publication appeared in the 1845 work *History of Kington* as a five-page essay titled, "An Essay on the Best Method of Conducting the Kington's Mechanic's Institute." Wallace recommended an emphasis upon science for the Institute and urged the collecting of serious scientific treatises over works of "too trifling a nature."[13] Interestingly, the essay further reveals that Wallace had abandoned Old Testament special creation in favor of an old earth conception of change operating through universal laws.[14] This is not to suggest that he was in any sense a transmutationist at this point, but the intellectual ground had prepared the way for its growth.

More broadly, Wallace's Neath years placed him within a very active intellectual and cultural community. The Kington Mechanics Institute had a growing library of nearly 4,000 volumes that contained Charles Lyell's *Principles of Geology*, Humboldt's *Personal Narrative of Travels*, and Darwin's *Journal of Researches* hot off the press; a Philosophical and Literary Society had been formed in 1834; and the Neath Museum contained an *Ichthyosaurus* along with specimens of British birds, a collection of fossils, geological minerals, and even a shell collection.[15] Wallace was active in the Institute lecturing on various scientific topics, served as curator for the Philosophical and Literary Society's museum, and attended meetings of the nearby Swansea Royal Society.[16] In fact, Wallace was so active with the Institute that it asked that he and his brother John design and supervise the construction of its building. Though not opened until Alfred had left for South America, it remains a permanent testament to his active interest in that community and in egalitarian educational institutions in general.

Unfortunately, there were less congenial aspects to Wallace's time at Neath. As a surveyor, his services were called upon to prepare the way for the Enclosure Acts, a government redistribution plan under the guise of land "improvement" that saw the displacement of thousands of rural residents. Alfred's role in this "all-embracing system of land-robbery" was a matter of extreme dissatisfaction recalled angrily years later and unquestionably paving the way for his support of the Land Nationalization movement.[17] Another distasteful duty he had to perform while at Neath was a survey for the purpose of a valuation of property in the Gnoll estate. When he completed the work he was then informed that he also had to collect payment from the farmers in the district. Wallace was put off by the whole "disagreeable business" since these poor, Welsh-speaking laborers were ill-equipped to assume any additional financial burden. The entire episode was sufficient to lead the young surveyor away from any and all business pursuits—a life in science was clearly beckoning.[18]

That life was yet to take a definite direction, but his Neath years can be seen as formative. Not only was Neath's lively cultural climate conducive to stimulating Wallace's scientific interests, interests only made keener by the more odious aspects of his surveying work, but the ideas that would animate his life's work are to be found here. Most interesting is a lecture delivered by a Dr. Thomas Williams "On the theories propounded in the Vestiges of Creation," a summary review and analysis of Robert Chambers's famous (some said "infamous") transmutationist treatise *Vestiges of the Natural History of Creation*, which Wallace asked his friend Bates's opinion of in a letter dated December 28, 1845.[19] A year before Williams had published *A Sketch of the Relations Which Subsist Between the Three Kingdoms of Nature*. The most interesting feature of his *Sketch* is his discussion of "gradation" and "continuity" suggesting that animal life may be traced to the "junction-point of the nucleated cell." The substance of his argument was the action of secondary causes toward the operation of a unity of nature in an overall *scala naturae*. Whether or not Wallace actually heard Williams' talk is not known since the Swansea records of the event apparently have not survived; he makes no mention of them in his autobiography (then again he left much out of his Neath period), but the Williams lecture may indeed have prompted his inquiry to Bates. The seeds of Wallace's later development of his own brand of evolutionary theory, harkening back to a *scala naturae* unity, may indeed have been planted by Williams.

But it was Wallace's ongoing correspondence with Bates that would form the immediate link to his future. A letter in the spring of 1846, for example, shows Wallace eagerly exchanging insect lists and keeping a journal with Bates. Bates even visited Wallace at Neath in the summer of the following year. With the "great problem of the origin of species... distinctly formulated in my mind," convinced that the account in *Vestiges* was "so far as it went, a true one," and further "dissatisfied with a mere local collection"[20] from which little was to be learned, Wallace and Bates began to feel the irresistible pull of travel and adventure. The immediate catalyst was a reading of American entomologist William H. Edwards's *Voyage up the Amazon* in 1847. By spring of the following year the two men were on their way to South America.

ALFRED R. WALLACE. 1848
(From a daguerreotype)

2.

Four Years in the Amazon Valley

From 1848 to 1852

By pooling their financial resources along with the help of Bates's father, the two would-be naturalist/explorers boarded the 192-ton ship *Mischief* for Pará, Brazil leaving Liverpool on April 20, 1848.[21] They arrived at their destination on May 26. The two young men faced considerable challenges. For one thing not many British explorers had been to the Amazon before. Darwin, of course, had already skirted the South American coast during his global traverse on the *Beagle* (December 27, 1831 through October 2, 1836), but the interior rain forest remained largely *terra incognita* for English explorers. A beginning was made in 1835 when the Royal Geographical Society commissioned a German, Robert Schomburgk, to investigate British Guiana. Encouraged by the results, in 1838 Schomburgk embarked upon a seven-month, 3,500 kilometer exploration of the Uraricoera River in northern Brazil.[22] But aside from this the early nineteenth century left South America largely bereft of British explorers and scientists who instead tended to concentrate on Asia and the Middle East (especially the Nile region). Yet there were British commercial interests in the region, and with the botanist/explorer Richard Spruce, with whom Wallace would develop a lifelong friendship, close on their heels, they probably didn't feel completely alone.

The most fortunate aspect of Wallace's Amazon expedition, however, was who he left behind in England to look after his interests: Samuel Stevens. Stevens has been described as the "natural history equivalent of an impresario,"[23] a collector of beetles and butterflies who was well connected to the specimen auctioning business through his brother. It was Stevens who would make or break this Amazonian adventure for Wallace. Stevens agreed to pay four pence per specimen. After Stevens's 20 percent commission and 5 percent for insurance and shipping, only three pence remained for the collector. Slotten calls this "a pittance, but the going rate at the time."[24] Thus, for Wallace and Bates, quantity was essential. Indeed the need for large, steady supplies of insects would eventually cause Wallace and Bates to go their separate ways.

Stevens became indispensable to Wallace throughout his journeys, both this one and his later expedition to the Malay Archipelago. Stevens kept Wallace's cash flowing by negotiating the best prices for the specimens preserved and shipped to him, and as his agent in England and as treasurer for the Entomological Society of London, Stevens brought the Wallace name into an important circle of curators and naturalists.[25] The down side was that Wallace was under constant pressure to keep large numbers of specimens to Stevens flowing.

It was costly in another way too. Edward, Alfred's younger brother, insisted on trying his hand at the naturalist/explorer's life but found the deprivations of tropical living too arduous to endure and left Wallace to return home. While at Pará, as his older brother explored the Uaupés River, he became ill. Even the procurement of the best medical attention available by Bates (also in Pará at the time) couldn't spare Edward from the lethal effects of yellow fever, and he died on June 8, 1851.

Meanwhile, Wallace knew none of this. Although Wallace had been up part of the "great unknown river"[26] before (in October of 1850), this second expedition up the Uaupés would take him deep into unexplored territory. The entire journey would last nine months, from June of 1851 to March of the following year. Of particular interest were Wallace's contacts with the many native tribes living along the banks of the Uaupés River and its tributaries. Wallace mentions 30 in all.[27] Unlike many British naturalists who dismissed or berated indigenous races, Wallace viewed the natives he encountered with intense interest. Giving details of their dress, domestic pursuits, crafts, religious beliefs, agriculture, dwellings, social life, etc., his descriptions are meticulous in ways that only a sympathetic eye could discern. Rejecting tales of Amazon warrior women as equivalent to "those of the wild man-monkeys,"[28] Wallace generally portrays Uaupés people as healthy, honest, and simple. The English explorer thought highly of mankind in this pristine state of nature and he bemoaned the effect that European influences would have upon them. Instead of being uplifted he thought they "will probably, before many years, be reduced to the condition of the other half-civilized Indians of the country, who seem to have lost the good qualities of savage life, and gained only the vices of civilisation."[29]

How different was Darwin's view of native peoples: His appraisal of the Fuegians was less than complementary. After comparing the women to mounds of hay, he thought little more of the men. Writing on December 25, 1832, his general assessment anticipated his *Descent of Man* by nearly forty years: "Viewing such men, one can hardly make one's self believe that they are fellow-creatures, and inhabitants of the same world. It is a common subject of conjecture what pleasure in life some of the lower animals can enjoy; how much more reasonably the same question may be asked with respect to these barbarians!"[30]

Much as Wallace appreciated the significance of being the first European in these uncharted lands among unknown peoples, the long and difficult trek up the Uaupés and Rio Negro took a toll on Wallace's health. He suffered bouts of dysentery and eventually contracted malaria. Sick and shivering with fever, he struggled back to Pará. Nonetheless, the experience was instructive. Although Darwin's explorations via the *Beagle* expedition and Huxley's voyage on the HMS *Rattlesnake* had already provided them with the raw data for their later evolutionary ideas, Wallace's excursions into the uncharted hinterland provided an intimate glimpse into worlds neither of his predecessors could appreciate. As Iain McCalman has astutely noted, "Wallace's four-year immersion in the gritty daily patterns of river travel had its compensations. It gave him an insider's perspective on both the habits and the lives of Amazon peoples.

He had far better opportunities to understand indigenous and foreign peoples than did his counterparts on British naval survey voyages. What Darwin and Huxley had merely sampled," he concludes, "Wallace experienced as part of everyday life. His interactions with locals of every stamp—whether Portuguese traders, half-castes, black slaves or 'half-wild' Indians—made him resistant to most European prejudices about empire and race."[31]

It was uniquely productive in other ways too. Contrasted with Darwin, who could afford to work at a more leisurely pace, Wallace had to keep a steady supply of specimens coming to Stevens; unlike Darwin, whose father fitted the bill for his *Beagle* voyage, Wallace's livelihood depended upon it. This influenced how he conducted his daily routine— normally a demanding 12-hours from dawn to dusk—and ultimately how he would construct his evolutionary theory. This will be covered in greater detail in Chapter 8.

But for now it was over. Tired, ill, and in desperate need of rest and recuperation, Wallace booked passage out of Pará on the *Helen*. He and an entourage of parrots, parakeets, macaws, other assorted birds, monkeys, dried and pressed plants, bird skins, sketches, notes, and maps boarded the 235-ton vessel on the morning of July 12, 1852. Wallace described his trip home as "rather adventurous," an understatement that fails to convey the disaster that awaited him. On August 6, at open sea, Captain Turner informed his passenger that the ship was on fire, a circumstance made all the more noxious by the inclusion of 120 tons of Indian rubber and a large store of palm oil on board. Wallace lost everything and very nearly lost his life. Too weak to hold the rope lowering himself onto the lifeboat, he stripped the flesh off his palms and for the next ten days sat miserably baking in the sun until rescued by the *Jordeson*. Finally, "*October* 1. Oh, glorious day! Here we are on shore at Deal, where the ship is at archor."[32] It had been an amazing four and a half years. Alfred Russel Wallace left England in the spring of 1848 an enthusiastic but inexperienced youthful adventurer; he arrived back in England in the fall of 1852 a field-tested explorer/naturalist.

3.

London Interlude

H IS ARRIVAL IN ENGLAND MUST HAVE BEEN QUITE A SIGHT. Disembarking with only the clothes on his back (a thin tropical shirt ill-suited to the cold October air), he had no home, no money, and little to show for his herculean efforts in the Amazon. Passersby probably thought him a pauper rather than an English naturalist. But Stevens proved more than an able agent, he provided aid and comfort when the battered traveler needed it most. With ankles swollen, legs ulcerated, and suffering from malnutrition, Wallace was nursed back to health by Stevens's mother. Just three days on land saw Wallace hobble into a meeting of the Entomological Society. Even better was the news from Stevens that he had insured Wallace's collection for £200; though £300 short of the estimated value, it was enough to tide the bedraggled wayfarer over for some time. Just two months later, in December, a recovered Wallace heard Thomas Henry Huxley give a talk (without notes) on the *Echinocci* parasite. Surprised to learn that Huxley was younger than himself, Wallace was impressed with Huxley's clear presentation delivered with rapid sketches on a blackboard (see the picture on page 60). Though he would come to disagree with the anatomist-not-yet-turned-Darwin's bulldog, his respect for the man's intellect and ability to exposit would always remain.

Still, Wallace had little to show for his efforts. Moreover, the object of his travels—the solution to the evolution question—continued to elude him. If he was to gain *anything* from his more than four-year ordeal he would need to convey the substance of his investigations to an English audience. He delivered a few papers to the Zoological and Entomological societies. One in particular, "On the Habits of the Butterflies of the Amazon Valley", delivered on November 7 and December 5 of 1853 to the Entomological Society hinted vaguely at transmutation. During this time Wallace spent considerable time at the British Museum and it is there that he met Darwin in the insect room where the two chatted, a meeting that left no impression on the Down House recluse.[33]

More urgent was getting his experiences in South America published. Taking his letters home, a few extant notebooks, and still vividly impressed memories, he published *Narrative of Travels on the Amazon* in 1853. That same year Wallace put together some of his surviving sketches and notes to publish *Palm Trees of the Amazon and Their Uses*. Together these two books have brought favorable comment from modern scholars, the *Narrative* still offering "compelling reading" for Iaian McCalman, and Ross A. Slotten referring to *Palm Trees* as "a major scientific contribution in economic or ethnobotany."[34] Unfortunately, the audience Wallace needed to impress did not agree. Darwin thought the *Narrative* lacked substance; even Wallace's friend, Richard Spruce, thought little of both slim volumes. Generally, Wallace's efforts were dismissed as the productions of a "mere collector." Nonetheless, Wallace ventured many ideas that he would later explicate more fully throughout the remainder of his career. The *Narrative* clearly shows that Wallace was thinking about the distribution of species. Correlating how species became diffused in adjacent areas from an original stock would require further distributional data, but Wallace's Amazon journey caused him to notice precise species boundaries in ways previous naturalists had not.[35] All in all, despite the costs to his health and his physical collections, the entire Amazon experience served as a valuable apprenticeship.

Yet Wallace returned to England having failed to do two things. First, entry into the inner circle of the British scientific community eluded him. Despite his extensive travels and large haul of specimens (about 10,000 not counting those that went down with the *Helen*), his two books were poorly received and his would-be peers viewed him as something of an upstart, a dealer in exotic goods for pecuniary gain unbecoming the gentlemanly aloofness from such crass commercialism demanded of and by scientific elites in Victorian society. The second thing that Wallace had failed to do, of course, is answer the question of species origins. To achieve success at both would require a second expedition.

At first, a return to the Amazon was contemplated. But realizing that his old collecting companion Bates was still in the region, he began to cast about for another fruitful region to explore. At meetings of the zoological and entomological societies that he attended with regularity plus visits to study the insect and bird collections at the British Museum, "I had obtained sufficient information," he recalled years later, "to satisfy me that the very finest field for an exploring and collecting naturalist was to be found in the great Malayan Archipelago, of which just sufficient was known to prove its wonderful richness, while no part of it, with the one exception of the island of Java, has been well explored as regards its natural history."[36] After some fits and starts at getting far eastern passage, he finally got passage (thanks to Sir Roderick Murchison) on the steamer *Bengal* heading for Singapore. On March 4, 1854, the Amazonian explorer turned eastward toward an adventure that would change his life and the history of biology.

NATIVE HOUSE, WOKAN, ARU ISLANDS
(Where I lived two weeks in March, 1859)

PORTRAIT OF DYAK YOUTH.

4.

The Central and Controlling
Incident of My Life

The Malay Archipelago, 1854–1858

REFLECTING ON THE EXPERIENCE NEARLY FIFTY YEARS LATER, WALlace called his work in the Malay Archipelago "the central and controlling incident of my life."[37] Indeed it was. It all started with his arrival at Singapore on April 20 after more than six weeks at sea. He stayed in Singapore several months collecting birds and bugs. He did not arrive alone. He had brought a sixteen-year-old boy, Charles Allen, with him as an assistant. He knew the lad as the son of a London carpenter who came recommended by his sister Fanny and his brother in-law Thomas Sims, but Charles could never quite meet the demanding naturalist's high expectations and Wallace noted that Charles didn't remain long (about 18 months) before gaining employment at one of the Singapore plantations. The schedule he gives of his daily activities while in Singapore is instructive in showing the general daily routine he would adopt throughout the remainder of his stay in the East:

> Get up at half-past five, bath, and coffee. Sit down to arrange and put away my insects of the day before, and set them in a safe place to dry. Charles mends our insect nets, fills our pin cushions, and gets ready for the day. Breakfast at eight; out to the jungle at nine. We have to walk about a quarter mile up a steep hill to reach it, and arrive dripping with perspiration. Then we wander about in the delightful shade along paths made by the Chinese wood-cutters till two or three in the afternoon, generally returning with fifty or sixty beetles, some very rare or beautiful, and perhaps a few butterflies. Change clothes and sit down to kill and pin insects, Charles doing the flies, wasps, and bugs; I do not trust him yet with beetles. Dinner at four, then at work again till

six: coffee. Then read or talk, or, if insects very numerous, work again till eight or nine.[38]

Dissatisfied with his catches in and around Singapore, Wallace and Charles went to Malacca for around nine weeks. Upon returning to Singapore Wallace became the local guest of the "White Rajah," Sir James Brooke, who had been in the Malay Archipelago for fifteen years and established over the years a rather benevolent dictatorship over the region. The rajah soon convinced Wallace to go to Sarawak in Borneo. He arrived on November 1, 1854, and remained the next fourteen months.[39] It was in Borneo that Wallace encountered, lived with, and grew rather fond of the Dyak headhunters in the region. One might think that native headhunters would repulse an English Victorian visitor, but not Wallace. "The old [Dyak] men here related with pride how many 'heads' they took in their youth," he noted with some sympathetic interest, "and although they all acknowledge the goodness of the present rajah [Brooke], yet they think that if they were allowed to take a few heads, as of old, they would have better crops. The more I see of uncivilized people," Wallace concluded, "the better I think of human nature on the whole, and the essential difference between civilized and savage man seem to disappear."[40] Wallace came to live with and implicitly trust the Dyaks. While the Englishman appreciated these people more than most of his fellow countrymen, certain ceremonies could become annoying. At one such event, for example, a local Orang Kaya (rich ruler) arrived to much fanfare: "All the time six or eight large Chinese gongs were being beaten by the vigorous arms of as many young men, producing such a deafening discord that I was glad to escape to the round house, where I slept very comfortable with half a dozen smoke-dried human skulls hanging over my head."[41] This incident is interesting in demonstrating the ease and mutual trust Wallace was able to foster with indigenous peoples who might have "offended" Victorian sensibilities. This fact alone would be an important feature distinguishing Wallace from all his fellow naturalists (Darwin, Huxley, and Hooker).

The inadequacies of Charles as a naturalist's assistant forced Wallace to cast about for a replacement, and he found one at Sarawak: a 14-year-old Malay boy named Ali who would remain the adventurer's constant aide and companion for the rest of his stay in the region.

If the Amazon was Wallace's apprenticeship in natural history, the Malay Archipelago was where his ideas began to come together. To facilitate this process he began a separate journal to include a hodgepodge of articles and scientific text extracts, observations, anecdotes, and assorted musings. Fascinated by the orangutan, he would publish an article on the human-like beasts in the *Annals and Magazine of Natural History*. But of much greater significance was an essay, "On the Law Which Has Regulated the Introduction of New Species" (first appearing in the *Annals and Magazine of Natural History* [September 1855]) that is often simply called Wallace's "Sarawak Law" paper. Wallace was in Sarawak from November 1, 1854 to January 25, 1856;[42] the paper was written in February 1855 during the rainy season, with only Ali as his cook and companion, Wallace had time to pour over his books and ponder his experiences. Putting his ruminations to paper, Wallace made a bold proposal comprised of nine "facts" forming one overarching law, namely, that four specific geographical principles and five geological principles suggested the following: *"Every species has come into existence coincident both in space and time with a pre-existing closely allied species* [italics in the original]." Here is a summary of its basic principles:

Wallace's Sarawak Law

Four geographical principles:

1. The distribution of large classes and orders is significant;
2. Distributionally distinctive genera are important;
3. Natural species affinities are almost always geographically circumscribed;
4. Countries with similar climate, though separated by wide seas or large mountains, will exhibit families, genera, and species closely allied to one another.

Five geological principles:

1. Distribution of the organic world in time is close to their distribution in space;
2. Most large, and a few smaller, groups extend through several geological periods;
3. Each geological period, however, includes unique groups not found elsewhere;
4. Species of one genera or family within a period are more closely allied than those from different periods;
5. The appearance of groups and species are singular events, no group or species has come into existence more than once.

The importance of Wallace's Sarawak Law paper has not been missed by subsequent scholars. Admitting that the actual mechanism *causing* evolutionary change remained to be discovered, the famous American paleontologist and geologist Henry Fairfield Osborn nonetheless declared Wallace's contribution "a very strong argument for the theory of descent, as explaining the facts of classification, of distribution, and of succession of species in geological time. Wallace," he called, "a strong and fearless evolutionist...."[43] More recently, Iain McCalman called the Sarawak paper "the first ever British scientific paper to claim that animals had descended from a common ancestor and then produced closely similar variations which evolved into distinct species."[44]

Unfortunately, Wallace's Sarawak Law article got about as much attention as did his books on the Amazon and palm trees. There were a couple of exceptions. Sending a copy to his old Amazon colleague Henry Walter Bates, Bates declared, "The idea is like truth itself, so simple and obvious that those who read it and understand it will be struck by its simplicity and yet it is perfectly original."[45] More importantly, Charles Lyell, the geologist who had so influenced both Wallace and Darwin and remained a Down House confidante, read Wallace's paper and was frankly rattled by its persuasively argued uniformitarian thesis against his own objections to evolution. He felt beat at his own game. When

Lyell met with Darwin in April 1856, he seriously urged the now preeminent expert in barnacles to publish his theory as soon as conceivably possible. Darwin read Wallace's paper too but claimed it contained "nothing very new."[46] It's hard to say why Darwin missed Wallace's point. Perhaps he was simply too self-absorbed to see it. His leading biographer suggests as much. "Usually so alert to the different ways of seeing nature, Darwin blindly stared past the implication in Wallace's words. Though looking outwards," Janet Browne admits, "he was not prepared to see the possibility that someone else might be hesitantly circling around before arriving at the same theory. His own work, not Wallace's, was primary."[47] Still, the overall silence which met Wallace's Sarawak effort remains inexplicable. Even Huxley, writing years later, noted, "On reading it afresh I have been astonished to recollect how small was the impression it made."[48] There was another group unimpressed with Wallace's publication. Writing to Samuel Stevens, his dutiful agent replied that "several naturalists" (perhaps including Stevens who needed specimens to keep the cash flow going) were disappointed by his "theorizing." What they wanted were specimens not speculations! Wallace needed to get back to work.

Perhaps his agent and his clients back in Britain were spoiled. By the end of January 1858, after Wallace had been in the Malay Archipelago nearly four years, he could write proudly to his entomologist friend Bates that he had collected 620 species of butterflies, 2,000 species of moths, 3,700 species of beetles, 750 bee and wasp species, 660 fly species, 500 species of "bugs, cicadas, etc.," 160 species of locusts, 110 species of dragonflies, and 40 earwig species: 8,540 species in all![49] This does not include numerous bird and mammal species.

Thanks to his indefatigable collecting, from both Wallace's and Stevens's perspectives the Malay venture had proven a commercial success. And for good reason—Wallace was traversing the entire archipelago. June and July 1856 he spent exploring and collecting on the islands of Bali and Lombok; in the fall of that year he landed on Macassar at the island of Celebes; and in 1857 he spent his time in the Aru Islands and the

Moluccas. This former group he found so delightful that he returned
there in 1859. Here he found one of the most exotic and prized creatures
he had longed to encounter: "New and interesting birds were continually
brought in," he recalled with relish, "either my own boys or by the natives,
and at the end of a week Ali arrived triumphant one afternoon with a
fine specimen of the Great Bird of Paradise. The ornamental plumes had
not yet attained their full growth, but the richness of their glossy orange
colouring, and the exquisite delicacy of the loosely waving feathers, were
unsurpassable."[50] He was perhaps the first Englishman to see this ex-
quisitely beautiful creature in its wild habitat. Wallace summarized his
entire time in the Aru Islands thus:

> My expedition to the Aru Islands had been eminently successful.
> Although I had been for months confined to the house by illness, and
> had lost much time by want of the means of locomotion, and by miss-
> ing the right season at the right place, I brought away with me more
> than nine thousand specimens of natural objects, of about sixteen hun-
> dred distinct species. I had made the acquaintance of a strange and
> little-known race of men; I had become familiar with the traders of the
> East; I had reveled in the delights of exploring a new fauna and flora,
> one of the most remarkable and most beautiful and least known in the
> world; and I had succeeded in the main object for which I had under-
> taken the journey—namely to obtain fine specimens of the magnificent
> Birds of Paradise, and to be enabled to observe them in their native
> forests. By this success I was stimulated to continue my researches in
> the Moluccas and New Guinea for nearly five years longer, and it is still
> the portion of my travels to which I look back with the most complete
> satisfaction.[51]

Wallace also spent time in the Moluccan islands: Banda, Amboyna,
Ternate, and Gilolo. Grand as his time in the Aru Islands was, his time
in this island group would be memorable. He arrived on the island of
Ternate on January 8, 1858, "the fourth of a row of fine conical volcanic
islands which skirt the west coast of the large and almost unknown island of Gilolo."[52] *Here* history would be made.

MY FAITHFUL MALAY BOY—ALI. 1855-1862

Charles Darwin about 1854

5.

DARWIN AND THE TERNATE LETTER

THEORIES SO CLOSE AND YET SO FAR

ONLY TWO THINGS COULD EVER GIVE A WORKING NATURALIST LIKE
Wallace time to theorize: bad weather or bad health. Bad weather
spawned his Sarawak Law essay, now poor health—malaria to be exact—would birth his evolutionary mechanism, the drive train of change.
During February of 1858, in between bouts of fever, he recalled Thomas
Malthus's *Essay on the Principle of Population*. Thinking over the struggle and constant destruction—Malthus's "constant checks"—on unbridled population growth, Wallace asked, "Why do some live? And the
answer was clearly that on the whole the best fitted live. From the effects
of disease the most healthy escaped; from enemies, the strongest, the
swiftest, or the most cunning; from famine, the best hunters or those
with the best digestion; and so on. Then it suddenly flashed upon me,"
he continued, "that this self-acting process would necessarily *improve the
race*, because in every generation the inferior would inevitably be killed
off and the superior would remain—that is, *the fittest would survive*."[53]
At least that's how he recalled it nearly a half century later. It should be
pointed out in his reminiscence, however, that Wallace applied variations on the phrase "survival of the fittest," something he actually borrowed from his friend Herbert Spencer and later suggested to Darwin
to include in his *Origin*, which he did in the fifth edition published in
1869.

Wallace spent the next three evenings writing out what became
known as his "Ternate letter." On March 9, 1858, Wallace sent the letter to Darwin, asking only that he share it with Lyell if he thought it
"sufficiently important."[54] On June 18 Darwin wrote in his journal: "in-

terrupted by letter from AR Wallace."[55] The receipt of the now-famous letter requires an important digression. Darwin's reaction and response provide a unique opportunity to glimpse the inner workings of the man who would become the "father of modern biology."

To begin with, Darwin was shaken, but he should have seen it coming at least since reading Wallace's Sarawak paper. Nonetheless, what Darwin simply couldn't see before now he *now* saw in bold relief—*his* theory was about to be usurped! It is hard to grasp the full impression that Wallace's Ternate letter must have had on the already neurotic and dyspeptic Victorian. Darwin was always unusually protective of his work, not the least of which was his evolving theory of species origin. "One of Charles Darwin's few character flaws was this," observes one recent analyst, "he was oddly possessive about *his* theory, so much so that he failed to acknowledge his predecessors, including his own grandfather [Erasmus Darwin's *Zoonomia*], until his detractors pointed out the glaring omissions. He wanted the theory of evolution to be *his* discovery, *his* creation, *his* baby. He was, to say the least, single-minded in the intensity of his devotion."[56]

Given his passionate attachment to his theory, a lesser man might have simply destroyed the letter and quickly unveiled his work, or at least some version of it, to the public. After all, the British mail system was operating at peak efficiency by the 1850s, but it wasn't flawless; it wouldn't have been the first or last letter ever to get "lost" especially one coming from the other side of the world. But Darwin was if nothing else a man very much concerned with propriety and image. The attempt to suppress Wallace's letter was full of risks. What if it could somehow be demonstrated that the letter *had* been delivered? What if Wallace made trouble? What would his peers think of his theory if sullied by even the implication of such a scandal? Above all, Darwin hated controversy and contention. The option of disregarding the Ternate letter was not one well suited to a man of his temperament *and* sense of fair play.

Yet how could Darwin have let this happen? The question of Darwin's delay remains, on the face of it, perplexing and looms large.[57] After all, he had read Malthus by 1838 and had the essential elements of his theory in place. By July 5, 1844, he had written out a 230-page draft that he was convinced would "be a considerable step in science," had it transcribed, and deposited with his wife Emma with instructions to forward to a short list of possible editors (Charles Lyell first heading the list later replaced with Hooker) and an offer of £400 to complete the work.[58] Janet Browne says of this 1844 sketch that Darwin proposed that transmutation occurred through "chance and change, and depended on intimate links between geology and biology; everything he ever expressed interest in during the *Beagle* voyage or while excitedly dashing off species notebooks fused together into one tightly woven scheme.... Though inadequately buttressed with examples, as he sadly acknowledged from the start, and lacking any considered analysis of the history of one group of islands which might substantiate some of his points, Darwin's essay of 1844 was a magnificent *tour de force*."[59] But if this *"tour de force"* was so "magnificent," why wait? Why not supply the omissions and correct the defects now and proceed to press?

A number of suggestions have been offered, but two stand out. First, Darwin himself wasn't ready. Darwin, for all his voyaging, still had yet to establish absolute mastery over an area of biology. An exchange between himself and Joseph Hooker apparently provided the impetus to do so and barnacles provided the specific means. Darwin was convinced that a thorough study of barnacles (Cirripedia) was called for. From 1846 to 1854 he did little else. Darwin found in his meticulous and painstaking studies homologies suffused with variation; where he once thought animals and plants more or less stable unless altered by some ecological circumstance, he now found "confounded variation, which, however, is pleasant to me as a speculatist [sic]...."[60] As Wallace's biographer Ross A. Slotten puts it, "Barnacles taught Darwin what Wallace would learn from his experiences as a field biologist: that variations occurred naturally and spontaneously."[61]

From 1851 to 1855 Darwin published an exhaustive account of fossil and living Cirripedia (see Marsha Richmond "Darwin's Study of the

Cirripedia" and "note on dates" in http://darwin-online.org.uk). He was awarded the prestigious Royal Society's Medal in 1853. Darwin was now a recognized authority and a respected biologist. But it came at a cost—an eight-year delay. Hooker even tired of the barnacles study and pleaded for some news of his broader theory, leading Darwin to quickly remind him that it was his "decided approval" of his barnacle work that had caused him to "defer my species paper" to begin with.[62]

This at least accounts for an eight-year interval between his 1844 sketch and final publication. Still, the additional five-year lag remains to be explained. Here one must conclude that Darwin was gripped by uncertainty and perhaps even fear over the public reaction to so full and disturbing an explication of evolutionary theory. His theory of evolution countered special creation—the idea that the universe was created out of nothing by a special act of God thousands rather than millions of years ago and that each life form on earth also represented a special act of creation by God with each uniquely adapted to their respective environments—by suggesting that the universe was a product of slow incremental change and that all biological life was the product of struggle driven by wholly random and chance forces (mainly through natural selection).

The implications for mankind were so disturbing to a society steeped in one version of the Genesis creation story that Darwin purposely avoided the topic of human evolution in *Origin*. But the disruption posed by the new philosophy of science that undergirded Darwin's theory went even further. Darwin's brand of evolution not only suggested blind forces operating to create what was once the purview of the sacred, it also was premised upon methodological naturalism, the notion that scientists *must* invoke *only* natural processes functioning via unbroken natural laws in nonteleological ways. Darwin's methodological naturalism was, in fact, the engine that drove an implicit commitment to a materialistic metaphysic, the sources of which were largely two-fold: first a rather worldly paternity that never took religion too seriously (Darwin's father Robert was at best a deist and his grandfather, Erasmus, wrote a two-volume work on transmutation titled *Zoonomia* first published in

1794, which young Charles read rather carefully) and second his early experiences as a failed teenage medical student at the University of Edinburgh.

Darwin's Edinburgh experience was especially transformative. While there he petitioned for entry into the Plinian Society, a loose-knit body of freethinkers, on November 21, 1826. Through the Plinians he would be exposed to some of the most heretical views of the day.[63] That very meeting heard William Browne, who had proposed the young Darwin (seventeen) for membership minutes before, deliver a talk countering Charles Bell's *Essays on the Anatomy of Expression* (1806), which insisted that the Creator had endowed the human face with a form and structure uniquely suited to the expression of human emotion. Browne thought it "anatomical chauvinism" to assume any special difference between animal and human facial anatomy. The next week Darwin heard William Greg, a fellow student and just as iconoclastic as Browne, give a presentation setting out to prove that "the lower animals possess every faculty & propensity of the human mind."[64] On March 27 Browne returned to give an inflammatory lecture on mind and matter. Browne told the astonished students that mind and consciousness were merely the result of brain activity. This was seen as so potentially dangerous that it was struck from the Society's minutes.[65]

A few heterodox lectures would hardly be enough to set an inquisitive but inherently conservative teenager along the path to heresy. But his relationship with Robert Edmond Grant was different. Fellow Plinian and sixteen years Darwin's senior, Grant was an expert on aquatic invertebrates who became Darwin's closest confidant. The budding naturalist would gain foundational knowledge under Grant's tutelage. But he would gain something else as well. "Theirs was a decisive meeting," write Adrian Desmond and James Moore, "Darwin was coming under the wing of an uncompromising evolutionist. Nothing was sacred for Grant. As a freethinker," they continue, "he saw no spiritual power behind nature's throne. The origin and evolution of life were due simply to physical and chemical forces, all obeying natural laws."[66]

So with a secular worldview no doubt learned at his father's knee and the Edinburgh Plinians well behind him, Darwin's experiences on the *Beagle* were set within a mental template of materialism. While this is not the view we get from his *Autobiography* (posthumously published in 1887), a careful perusal of his private notebooks shows that even in its earliest development Darwin's evolutionary theory was founded upon materialistic assumptions. Howard E. Gruber frankly admits that "Darwin presented himself in ways that are not supported by the evidence of the notebooks," and his "actual way of working... would never have passed muster in a methodological court of inquiry among Darwin's scientific contemporaries."[67] Gruber concludes that Darwin amassed much of his purported evidence "after his views were quite well developed."[68] In short, Darwin the tireless, unbiased investigator patiently following accumulating pieces of data wherever they may lead is a fiction largely perpetrated by Darwin himself. It has been summarized best by historian of science Stanley Jaki:

> In writing his autobiography Darwin did not recall the delight he experienced as he perceived in the course of filling his first *Notebooks* that if his evolutionary theory were correct and if "conjecture" was allowed "to run wild," then we—animals and humans—"may be all melted together!" Much less would he have been willing to recall that this "melting of all together" in a purposeless flux and especially the "melting down" of man into just another species was his chief inspiration from almost the moment he stepped off the *Beagle*. The publication in full of Darwin's *Early Notebooks* [individual notebooks were published in the 1970s with the complete compilation issued by the British Museum and Cambridge University Press in 1987] forces one to conclude that in writing his *Autobiography* Darwin consciously lied when he claimed that he had slowly, unconsciously slipped into agnosticism. He tried to protect his own family as well as the Victorian public from the shock of discovering that his *Notebooks* resounded with militant materialism. The chief target of the *Notebooks* is man's mind, the "citadel," in Darwin's words, which was to be conquered by his evolutionary theory if its materialism were to be victorious.[69]

Silvan Schweber essentially agrees with Jaki. "I believe that by July 1839 Darwin had a unitary evolutionary view of everything around him: the planetary system, our own planet, its geology, its climate, its living organisms and their social organizations. More important, he had convinced himself that the mechanism of this evolutionary process was accounted for by the invariable laws of physics and chemistry and the principle of natural selection, *without* the necessity of divine intervention at *any* stage or level. The acquisition of this vision from 1837 to 1839 entailed profound religious consequences, and the M and N notebooks and the Old and Useless notes are also an account of Darwin's search for God. He did not find Him, and by 1839 Darwin was certainly an agnostic (and possibly an atheist)."[70]

It is important to realize that the sources—call them the *seeds*, if you will—of Darwin's materialism and his religious abandonment were sown much earlier, at least during the nineteen Plinian Society meetings he attended between November 21, 1826 and April 3, 1827. From that point forward Darwin's march towards materialism might be called ineluctable, and from around 1839 (perhaps as early as the fall of 1838[71]) a *fait accompli.*

But Darwin's rhetorical style often obscures his philosophical commitments. These intellectual antecedents compelled Darwin to cautiously craft his evolutionary theory by inventing "a phrase poised on the edge of metaphor," writes Gillian Beer, "a phrase that, moreover, alluded to its predecessor, even as it undermined it: 'natural selection' is a pithy rejoinder to 'natural theology'. Instead of an initiating godhead, Darwin suggests, diversification and selection have generated the history of the present world. Instead of teleology and forward plan, the future is an uncontrollable welter of possibilities. In the world he proposed there was no crucial explanatory function for God, nor indeed was there any special place assigned to the human in his argument."[72] Herein lays Darwin's brilliance: through analogy, metaphor, double entendre, rhetorical sleight of hand, topical construction, and careful psychological reconditioning of the reader into his world, he was able take readers of *Origin* from Paley's world into his own. All truth claims aside, it was a rhetorical *tour de force*.[73]

Whatever strategies Darwin employed, they always implied, and indeed were drawn from, an overarching commitment to materialism whose operational logic found place in naturalism. Darwin did not take these philosophical commitments lightly. In fact their implications weighed heavily upon him. In a letter to Joseph Hooker on January 11, 1844, he admitted, "it is like confessing a murder," a murder of English society's most cherished beliefs, beliefs that provided the moral glue holding civil society together, and even the murder of God himself.[74]

Later that year he wrote to the Anglican priest/naturalist Leonard Jenyns, "I know how much I open myself to reproach from such a conclusion [that species are mutable and descend from common stocks], but I have at least honestly and deliberately come to it. I shall not publish on this subject for several years."[75] Well, if not *entirely* "honestly" it was at least honestly deliberate. The linkage of "reproach" with delay is unmistakable. Yet it was a belief nonetheless adamantly—indeed *faithfully*—held.

A few years later, in 1848, Darwin wrote to a resistant Hooker revealingly, "I don't care what you say, my species theory is all gospel."[76] Even on the eve of final publication, Darwin engaged in some fretful handwringing. Writing to Lyell he asked whether or not he should "tell Murray [John Murray, his publisher] that my book is not more *un*-orthodox than the subject makes inevitable. That I do not discuss the origin of man. That I do not bring in any discussion of Genesis, & c., & c., and only give facts, and such conclusions from them as seem to me fair. Or had I better say *nothing* to Murray, and assume that he cannot object to this much unorthodoxy," adding with a none too subtle reminder of his own friend's challenge to what was then perceived as the geological catastrophism mandated by Genesis with his own gradualist uniformitarianism in his *Principles of Geology*, "which in fact is not more than any Geological Treatise which runs slap counter to Genesis."[77]

Darwin was a conflicted man; literally sick with concern for priority and sick with worry over its consequences. Meanwhile, a very sick Wallace had just sent him a bombshell. The arrival of his letter titled "On the tendency of Varieties to depart indefinitely from the Original Type" (see Appendix A) brought into convergence at least the perception of one mechanism—natural selection—and two very different men. The one a seasoned naturalist/explorer whose philosophy, still inchoate, was at this point little more than an Owenite idealism so befitting of a tradesman educated at a mechanic's institute; the other an independently wealthy Cambridge graduate and expert on barnacles whose materialistic philosophy was as fixed and immutable as the species of special creation he sought to supplant. This convergence prompted action.

What to do? Do nothing and risk preemption; rush to press as if nothing happened and risk controversy or worse—scandal. Darwin asked his close friends and confidantes Joseph Hooker and Charles Lyell for a solution.[78] It couldn't have happened at a worse time, for Charles and Emma had serious illness in the house. Both 15-year-old Henrietta and nineteen-month-old Charles Waring Darwin were down with raging fevers. Henrietta would survive, but nearly a week later (on June 28) the baby Charlie would die of scarlet fever.

Amidst these consuming distractions, the senior Charles would rely completely upon the collective judgments of his old friends Hooker and Lyell. They both recommended jointly presenting Wallace's letter along with material on Darwin's theory to the Linnean Society. Time was of the essence since Lyell and Hooker had no idea if Wallace had a copy or perhaps another version floating around elsewhere. Since Darwin, Lyell, and Hooker were all fellows of the Society, getting the papers into the next meeting would be no problem. It would be a boost for Wallace since he had no formal connections to the organization. Although Wallace was known to its members (especially William Wilson Saunders, a well-heeled entomologist collector who was one of Wallace's regular customers), his exclusion from the group was probably the product of class-conscious snobbery. Desmond and Moore are probably correct in characterizing the impression Wallace left upon most Society members as that of a "specimen haggler."[79] In any case, Wallace was lucky "to hitch a ride on Darwin's well-cut coat-tails."[80] In the end, Darwin cobbled together extracts of his 1844 sketch (important in confirming Hooker's reading of it shortly after completion and leaving no doubt as to priority for Darwin) and an 1857 letter to the famous American botanist Asa Gray (essential because it included material learned from his barnacle work that superseded the sketch). Lyell and Hooker dropped the material off to the Linnean Society secretary on June 30, 1858. The theory of evolution was formally unveiled the following evening.

The actual reading of the papers was anticlimactic (see the proceedings). Few members (about 30 in all) showed at Burlington House. First came Darwin's 1844 extract, then his letter to Asa Gray, and finally Wallace's Ternate letter, "On the tendency of Varieties to depart indefinitely from the Original Type." Sadly, Darwin had a baby to bury and couldn't attend. The entire event was engineered to establish not only Darwin's priority but his preeminence as well. "Even Darwin winced," notes Browne, "when he saw the layout some weeks later."[81]

No one seemed to recall the meeting very well and Wallace's paper, last to be read, was certainly lost upon the bleary-eyed audience. The important point of the meeting was two-fold: first, it firmly established Darwin's priority; and second, the audience (including Darwin) *thought* they heard the same theory. As will be discussed later, they were wrong. In any case, the entire affair was forgettable enough to prompt Linnean Society President, Thomas Bell, to comment in his 1859 address, "The year which has passed, has not, indeed, been marked by any of those striking discoveries which at once revolutionize, so to speak, the department of science on which they bear."[82]

Despite the lack of impression made that July 1st evening at the Linnean Society's hastily gathered meeting, in retrospect Wallace's Ternate letter was, by several assessments, "the more impressive."[83] Examinations of all three pieces reveal Darwin's sketch and letter to Gray full of the tentative uncertainties that typified his writing. Halting and speculative, one historian has called Darwin's method (a method from which he would not retreat throughout his career) a "logic of possibility" in which "possibilities were promoted into probabilities, and probabilities into certainties, so ignorance itself was raised to a position only once removed from certain knowledge."[84] Although given the nature of their macroevoutionary speculations, Wallace could not entirely avoid this "logic of possibility" either, his writing was at least by contrast more coherent and confident. Even Darwin admitted, with unintended irony, "It puts my extracts... , which I must say in apology were never for an instant intended for publication, in the shade."[85]

After the event Darwin and Hooker quickly sent letters to Wallace explaining what they had done. Ever magnanimous, Wallace's recollection a half century later reflects an assessment of his relationship to Darwin's theory that the persistent conspiracy theorists[86] would do well to consider: "Both Darwin and Dr. Hooker wrote to me in the most kind and courteous manner, informing me of what had been done, of which they hoped I would approve. Of course I not only approved, but felt that they had given me more honour and credit than I deserved, by putting my sudden intuition—hastily written and immediately sent off for the opinion of Darwin and Lyell—on the same level with the prolonged labours of Darwin, who had reached the same point twenty years before me...."[87]

With the Ternate letter now public and this all-important remote encounter behind them, and with Wallace cordially and supportively replying by post that it would have pained him had Darwin not "made public my paper unaccompanied by his own,"[88] Darwin could begin working feverishly to complete his book manuscript at least without the heavy burden of an absolute priority controversy looming. However neither Darwin nor Lyell nor Hooker knew for sure whether or not Wallace had a book manuscript tucked away in the wilds of some strange tropical island somewhere between Southeast Asia and Australia. Darwin would *still* need to rush his book to press.

John Murray released *On the Origin of Species* on November 24, 1859, not because he liked it or was even remotely convinced by it (he thought the theory "as absurd as though one should contemplate a fruitful union between a poker and a rabbit,"[89] he once remarked) but because he thought it would sell well. He was right. The book immediately sold out its initial print run of 1,250 copies and Murray took orders for 250 more. Mudie's Circulating Library's order for 500 guaranteed Darwin's book would get into the hands of a wide and diffused public. On December 1st Murray was already setting up for a new edition.[90]

The next edition Murray doubled to a print run of 3,000. Book sales would be driven by a proliferation of inexpensive review journals and magazines. *Household Words, All the Year Round, Cornhill Magazine, Chambers's Journal, Penny Magazine,* and even publishing house periodicals helping to create a market for the book trade like *Macmillan's Magazine* and *Harper's Bazaar* ensured that Darwin's theory would expand well beyond the stuffy high society dining halls and drawing rooms and into the homes of people Wallace was far more familiar with than Darwin.[91]

Meanwhile Wallace had unfinished business on the other side of the earth.

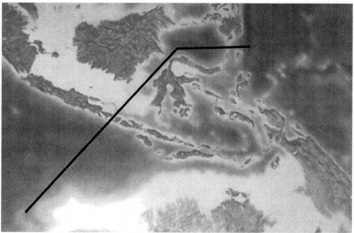

The Wallace Line Drawn Across Southeastern Asia

6.

UNFINISHED BUSINESS IN THE MALAY ARCHIPELAGO

LIFE ON THE ISLANDS AFTER TERNATE, 1858–1862

WALLACE WAS STILL IN THE MALAY ARCHIPELAGO INVESTIGATing the largely unknown islands of Ceram and its smaller sister to the south, Amboyna. While Darwin nervously awaited reaction to his species book near the throne of British Empire, Wallace was living with natives, missionaries, and an eccentric well-traveled Captain Van der Beck at Hatosúa, Ceram. The two co-discoverers were in dramatically different circumstances: Darwin in the posh comfort of Down House; Wallace finding few specimens, instead himself becoming a meal for nasty mites that left him "covered from head to foot with inflamed lumps."[92]

It had been a difficult time at Ceram and Amboyna but he was about to resolve a question that had perplexed him for years.[93] While on the islands of Bali and Lombok in the summer of 1856 Wallace took notice of the distinctive birds. At Bali he observed the Asian Golden Weaver, a bird at the extreme eastern limits of its range. Sailing to Lombok just 25 kilometers away he saw an entirely different bird, a Helmeted Friarbird, a relative of friarbirds found in Australia. Curiously the eastern birds he had seen on Bali were absent. In fact, Australian type birds abounded at Lombok, none of which could be found on Bali. Lyell had taught Wallace that the divisions of animal infraclasses and species needed large barriers—wide oceans, high mountains, sharp climatic difference—to work. But here was Wallace a mere 25 kilometers from where he had seen very different bird groups. Later, at Celebes, he noted similar patterns of animal life:

Here again we have a resemblance to the Wart-hogs of Africa, whose upper canines grow outwards and curve up so as to form a transition from the usual mode of growth to that of the Babirusa. In other respects there seems no affinity between these animals, and the Babirusa stands completely isolated, having no resemblance to the pigs of any other part of the world. It is found all over Celebes and in the Sula islands, and also in Bouru, the only spot beyond the Celebes group to which it extends; and which island also shows some affinity to the Sula islands in its birds, indicating perhaps a closer connexion between them at some former period than now exists.

The other terrestrial mammals of Celebes are, five species of squirrels, which are all distinct from those of Java and Borneo, and mark the furthest eastward range of the genus in the tropics; and two of Eastern opossums (Cuscus), which are different from those of the Moluccas, and mark the furthest westward extension of this genus and of the Marsupial order. Thus we see that the Mammalia of Celebes are no less individual and remarkable than the birds, since three of the largest and most interesting species have no near allies in surrounding countries, but seem vaguely to indicate a relation to the African continent.[94]

Finally, in January 1858 he wrote to his old South American compatriot Henry Walter Bates: "In the Archipelago there are two distinct faunas rigidly circumscribed, which differ as much as those of South America and Africa and more than those of Europe and North America. Yet," he added, "there is nothing on the map or on the face of the islands to mark their limits. The boundary line often passes between islands closer than others in the same group."[95] Wallace wrote up his findings more formally and sent it on to Darwin who dutifully forwarded it on for reading at a meeting of the Linnean Society on November 3, 1859, just weeks before Darwin would unveil his *Origin*. Published the following year as an essay "On the Zoological Geography of the Malay Archipelago", Wallace's article demonstrated that the archipelago was cut in two halves that "belong to regions more distinct and contrasted than any other of the great zoological divisions of the globe."[96] With the eye of an old surveyor Wallace had discerned distinct ecozones. Lyell had been wrong, massive barriers were *not* necessary to create distinct boundaries between species. Wallace opposed the notion of transoceanic migration of species and urged ancient land bridges (essentially a position of trans-continental land bridge extensions) and a far more malleable geographic landscape than once thought. Furthermore, hidden conditions rather than mere distance needed to be considered in looking at the distribution of flora and fauna. The deep but narrow straight between Bali and Lombok is an example. Wallace came to propose that the depths of sea divisions would be a more reliable test for similarities or dissimilarities between faunal groups. He abandoned his previous extensionist position because it argued against his more fixed theory of large and distinct "zoogeographical regions."[97]

Wallace would develop the insights gained in the Malay Archipelago and later broaden the distributional map developed by Philip Sclater in 1857 and apply it to larger categories of animals. For this he is often regarded as the father of biogeography. But the most noteworthy feature of biogeography, the one that would bear his name, was not established until Wallace had returned to England and drew a red line from the Makassar Strait (dividing Borneo and Celebes) joining the Celebes and Java seas on a map in an 1863 paper read to the Royal Geographical Society; to the west he labeled "Indo-Malayan region," to the east "Australo-Malayan region." The Wallace Line, "the most famous and most discussed boundary in the world," was born.[98] (See the map on page 48).

Despite having his Ternate paper read before one of the most prestigious scientific societies in England and his relationship with Darwin secure, including his key insights into biogeography, Wallace was not ready to go home. He was still on the prowl for more elusive birds of paradise. He traveled to the Ke Islands, Goram, headed back to Ceram, Waigiou (where he "brought away... twenty-four fine specimens of the *Paradisea rubra*," the rare red bird of paradise).[99] From Waigiou he returned to Ternate and on January 12, 1861, arrived at Delli, the capital of the Portuguese possessions in Timor. Spending some time in New Guinea, Wallace eventually took a steamer to Sumatra and from there went to Singapore. While there he found two healthy male birds-of-paradise, purchased them at a high price of £92, and resolved to take them back to England. At Singapore he left his trusted servant Ali his two guns, tools, and other items he thought might be of use. The somber visage (see the picture on page 35) from the photograph Wallace commissioned may have been the impending bittersweet farewell Ali knew awaited them both. The young man returned to Ternate wealthy (by Ternate standards), where he proceeded to vanish into history.

Heading back by way of Bombay and then to Egypt, Wallace spent most of his time tending and feeding his rather large menagerie of birds. Finally, he arrived back in England on March 31, 1862. Unlike his return from the Amazon, Wallace had much to show for his protracted adventures. Wallace's collection amassed during his eight years in the Malay Archipelago was nothing short of staggering: 300 specimens of mammals, 100 reptiles, 8,050 birds, 7,500 shells, 13,100 butterflies, 83,200 beetles, and 13,400 "other insects." A total of 125,660 in all![100]

Alfred Wallace in 1862

7.

RETURN TO LONDON

THE DOMESTICATION OF ALFRED RUSSEL WALLACE

WALLACE RETURNED HOME A FAMOUS BUT NOT A WEALTHY MAN. Fortunately he received a nice profit for his two birds of paradise: £150.[101] A more pressing question for Wallace was the reception he would receive in scientific circles. His election twelve days before his arrival in England as a fellow of the Zoological Society wasn't a bad start. Because of the Ternate letter, he was more than a nodding acquaintance with Darwin and his inner circle: Lyell, Hooker, and Huxley (who had taken no active role in setting up the unveiling of natural selection but by now was very familiar with Wallace's work).

Moving in with his brother in-law Thomas Sims and his sister Fanny in London, Wallace spent most of his time organizing his collections.[102] Nevertheless, Wallace needed to make a living. His most immediate option was to write, which he did with some ferocity. From 1862 to 1865 he authored twenty-eight papers on a range of subjects.[103]

One area that particularly interested him was comparative ethnology. His experiences spending some dozen years with various native peoples from South America to the Far East led him to some rather controversial conclusions regarding the nature of man. These views were framed by a belief (acquired from his days at the Mechanic's Institute as a young man) in phrenology, the idea first developed by Viennese physician Joseph Gall. Phrenology argued that character and even health could be discerned through the interpretation of bumps on the head and shape of the skull. Hardly a mystical science, Gall's theory laid to rest the ancient Roman physician Galen's association of the brain as the seat of "animal spirits" and Aristotle's earlier notion of *sensorum com-*

mune ("organ of the soul"). While the principles of phrenology did not ultimately prove viable, Gall was not the "sorry charlatan" his dismissed assistant Sourzheim tried to make of him, and historians of science now generally agree that much of Gall's more reliable work formed the basis for modern neurology.[104] Gall was correct in believing the brain to be the seat (though not necessarily the *source*) of all human faculties; he was wrong in asserting that they were topologic.

More importantly, phrenologists and its proponents (and there were many) believed that this progressive science could be used in the advance of society. Such an emphasis appealed to a man still devoted to the idealistic utopianism of Robert Owen. On March 1, 1864, Wallace read "The Origin of Human Races and the Antiquity of Man Deduced from the Theory of 'Natural Selection'" to the Anthropological Society. Wallace made a radical proposition, namely, that the normal operation of natural selection had been checked in man. Wallace argued that in the animal world survival was ensured by attributes—swiftness, stealth, strength, etc.—cultivated and culled over time. But how could one account for altruism and cooperativeness in humans? Wallace explained that in animals natural selection "keeps all up to a uniform standard":

> But in man, as we now behold him, this is different. He is social and sympathetic. In the rudest tribes the sick are assisted at least with food; less robust health and vigour than the average does not entail death. Neither does the want of perfect limbs or other organs produce the same effects in animals. Some division of labour takes place; the swiftest hunt, the less active fish, or gather fruits; food is to some extent exchanged or divided. The action of natural selection is therefore checked; the weaker, the dwarfish, those of less active limbs, or less piercing eyesight, do not suffer the extreme penalty which falls upon animals so defective.[105]

Wallace's unique analysis could have only come from a man intimately acquainted with humanity in a "state of nature." As Darwin's dismissive—indeed distraught—attitude toward the Fuegians demonstrated, he was not suited by class or circumstance to see the cooperative aspects of indigenous cultures. Viewed through his English lens of class and privilege, Darwin's only benchmark was the degree to which a given group diverged from his own standards of propriety; the degree to which they departed was the degree to which they became, in his eyes, brutes and beasts.

Nonetheless, although James Hunt, a founder of the Anthropological Society and white supremacist who believed the polygenist theory of separate ancestral descents for the different races, objected loudly to the common descent (monogenist) implications of Wallace's presentation, Darwin applauded Wallace's "great leading idea."[106] Hooker was "amazed at its excellence."[107] Darwin was enthusiastic regarding the paper for two reasons: first, at least at this point, Wallace simply seemed to be transferring the operations of natural selection from the human body to the human mind, a shift that retained Darwin's philosophical materialism; but secondly, Darwin was beginning to have doubts himself about the ability of natural selection to explain every aspect of the biological world and began to increasingly talk about sexual selection as an important subsidiary factor in biological development.

Lyell, with whom Wallace was developing a close mentor/protégé relationship, also warmly approved of the essay. More importantly, it is clear that Wallace echoed support of Charles Lyell's publication the year before, *Geological Evidences of the Antiquity of Man.* Lyell, never a firm or committed supporter of Darwin's evolutionary theory, wrote, "The whole course of nature may be the material embodiment of a preconcerted arrangement; and if the succession of events be explained by transmutation," he added, "the perpetual adaptation of the organic world to new conditions leaves the argument in favor of design, and therefore of a designer, as valid as ever.... It may be said that, so far from having a materialistic tendency, the supposed introduction into the earth at successive geological periods of life,—sensation,—instinct,—the intelligence of higher mammalians bordering on reason,—and, lastly, the improvable reason of Man himself, presents us with a picture of the ever-increasing dominion of mind over matter."[108] Though Wallace made no explicit reference to Lyell's words, they were impressed upon his own mind as he contemplated the minds of his fellow humans.

Meanwhile the practical exigencies of life remained. Wallace applied and was interviewed for a post with the Geological Society and was passed over twice. This disappointment was relieved with the return late in 1864 of his old friend Richard Spruce from Peru. Spruce was staying with a pharmacist and amateur bryologist named William Mitten. William's eighteen-year-old daughter, Annie, was a botany enthusiast and twenty-three years younger than Alfred. Despite the age difference, the two formed a close, insoluble bond. They were married on April 5, 1866. Together they would have three children: Herbert Spencer Wallace "Bertie" born June 22, 1867; Violet Wallace born January 25, 1869; and William "Will" Wallace born December 31, 1871.

As for a livelihood to support this growing family, writing remained his most immediate means of income. His most likely means of doing this would be to whip his notes and sketches of his eight-year adventure in the Malay Archipelago into shape for publication. Spending most of 1867 and 1868 in this pursuit, his book *The Malay Archipelago* was published in March of the following year. Receiving a £100 advance and generous royalty terms, this work more than any other would provide him with the most reliable income. Strong and steady sales prompted Macmillan to send the popular title through ten editions, the last in 1890. One of the most thorough and engaging voyage and travel narratives in the English language, it said to have influenced Joseph Conrad.[109] It is considered the most important book ever written on the region.

This was a period of dramatic change for a naturalist approaching middle age. While Wallace was courting Annie, he was persuaded by his sister Fanny to attend a séance on July 22, 1865. Investigating further, Wallace became a convert, along with a host of his scientific contemporaries (e.g., physicist Sir William Crookes, Columbia University's logician James Hervey Hyslop, physicist Sir Oliver Joseph Lodge, Nobel laureates Lord Rayleigh and Charles Richet, along with famed American philosopher/psychologist William James). "Wallace," observes biographer Peter Raby, "if not swallowing everything, was prepared to taste again and again."[110] His *Scientific Aspect of the Supernatural* (1866) was an explicit call for "an experimental enquiry by men of science into the alleged powers of clairvoyants and mediums."[111]

Wallace argued (contra Hume) that the so-called "supernatural" did not require by definition suspension or intervention contravening natural law but could, in fact, be the result of some "yet undiscovered natural law." Noting that much in the history of science once deemed "miraculous" is now attributable to one or more known laws, he proceeded to make a case for forces or entities operating on the physical world and that any *a priori* rejection of such a claim was "utterly gratuitous."[112] Wallace merely wanted a fair and unbiased critical assessment of spiritualist phenomena, calling upon the scientific community to at least take the anecdotal claims of their reputable and trustworthy colleagues seriously. It was a call from which he would never retreat. "We ask our readers not for *belief*," he pleaded, "but for doubt of their own infallibility on this question; we ask for inquiry and patient experiment before hastily concluding that we are, all of us, mere dupes and idiots as regards a subject to which we have devoted our best mental faculties and powers of observation for many years."[113]

It is important to note here that it is mistaken to simply dismiss Wallace's captivation with spiritualism as the product of a "heretic personality."[114] Despite widespread charlatanism, spiritualist advocates were not all kooks embracing the quirky, and they were not all naives and fools. Historians have struggled to understand spiritualism as a social phenomenon. While the frequent attribution of its popularity among leading Victorian intellects (especially scientists) to a "crisis of faith" no doubt has some important truth, it has also been pointed out that spiritualism was never completely refuted on its own terms. Stage conjurors (magicians) and leading opponents from the scientific community all failed to catch the age's leading spiritualist, Daniel Dunglas Home, in any chicanery. In the end, scientists who rejected spiritualism simply explained it away as wholly subjective.[115] This is not to vindicate spiritualism; it is merely to suggest that the movement posed a significant problem for scientists attempting to establish a meaningful discourse of objectivity and a normative basis for scientific inquiry based upon quantifiable empirical data on the one hand and those reliant upon personal testimony based upon experience and observation on the other. Seen in this light the contention over spiritualism is better viewed as an effort to negotiate precisely what counted as legitimate evidence rather than as a collection of aberrant eccentrics who merely provide historians comic relief in the otherwise serious and steady march of scientific progress.

As the decade drew to a close Wallace saw many changes. He started the decade in the wilds of the Malaysian islands; he ended it well domesticated with a son and a newly arrived baby daughter. Settled into a series of modest but comfortable homes, by 1869 Wallace and Annie's growing family remained in London but they began contemplating a move to the country. But appearances could be deceiving. Just as Wallace seemed to be settling down in his personal life, his professional life was about to take a dramatic turn.

Between his return to England in the spring of 1862 and the spring of 1869 all the elements for a major schism with Darwin and at least one of his evolutionary captains—what Iain McCalman has aptly characterized as *Darwin's Armada* —were in place. Up until now Wallace had been sailing triumphantly with the evolutionary squadron along with Captains Hooker and Huxley (and to a lesser extent Lyell) with, of course, Darwin as the admiral. But mutiny was afoot, and the occasion was a murder.

THE

MALAY ARCHIPELAGO

THE LAND OF THE

ORANG-UTAN AND THE BIRD OF PARADISE

A NARRATIVE OF TRAVEL

WITH STUDIES OF MAN AND NATURE

BY

ALFRED RUSSEL WALLACE

AUTHOR OF "DARWINISM," "ISLAND LIFE," ETC.

London

MACMILLAN AND CO.

AND NEW YORK

1890

The Right of Translation and Reproduction is Reserved

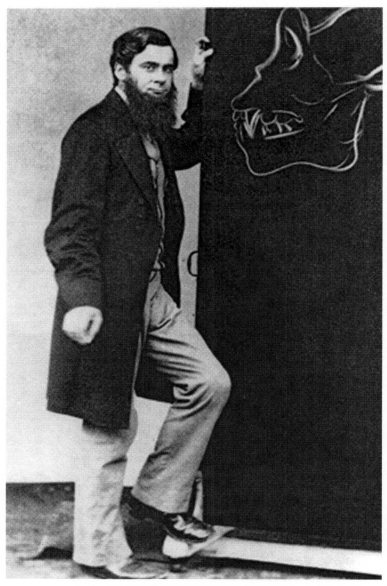

Thomas Henry Huxley in 1870

8.

MURDERING DARWIN'S CHILD

TOWARD AN INTELLIGENT EVOLUTION
AND A CLASH OF WORLDVIEWS

D ARWIN KNEW SOMETHING OMINOUS FROM WALLACE WAS IN THE
air. Writing to Wallace in March 1869, Darwin penned nervously,
"I shall be intensely curious to read the *Quarterly*: I hope you have not
murdered too completely your own and my child."[116] Darwin didn't have
long to wait for his "murder."

It came in the form of a review published in the April 1869 issue of
the *Quarterly Review*. It was Wallace's review of Charles Lyell's tenth
edition of his *Principles of Geology*.[117] Wallace and Lyell had established
a long and intense dialogue over evolution and the two agreed that the
theory—at least as Darwin had expounded it—carried certain implications for human development that were problematic; both became
sounding boards for each other regarding a teleological interpretation
of these processes.

Perhaps emboldened by his fertile discussions with Lyell, Wallace
used his review to, in Martin Fichman's words, present "to the world
the unambivalent evolutionary teleology that he would expound in
ever greater detail during the remainder of his life."[118] Wallace basi-
cally pointed to the human intellect as being too great for that simply
allowable by natural selection because, by definition, the law of natural
selection guided by the principle of utility (the idea that "no organ or at-
tribute can exist in a natural species unless it is or has been useful to the
organisms that possess it...."[119]) would be an effective barrier to its development. One could not, Wallace argued, explain the uniquely human
attributes of abstract reasoning, mathematical ability, wit, love of mu-

sic and musical aptitude, art appreciation and artistic talent, and moral sense as necessary for survival in a state of pure nature through which (by Darwin's own principle) natural selection must operate. Therefore, some other cause or action must be invoked. That cause of action Wallace called "an Overruling Intelligence."[120]

Darwin was devastated and scratched an emphatic "NO!!!" in the margin of his copy of the *Quarterly*. He wrote back to Wallace, "I presume that your remarks on Man are those to which you alluded in your note. If you had not told me I should have thought that they had been added by someone else. As you expected, I differ grievously from you, and I am very sorry for it."[121] Nine months later Darwin was still reminding Wallace, "But I groan over Man—you write like a metamorphosed (in retrograde direction) naturalist, and you the author of the best paper ["On the Origin of Human Races and the Antiquity of Man"] that ever appeared in the *Anthropological Review*! Eheu! Eheu! Eheu!—Your miserable friend, C. Darwin."[122]

Darwin also broached his disappointment to Lyell. Darwin did not get the sympathetic ear he was looking for. "I rather hail Wallace's suggestion that there may be a Supreme Will and Power which may not abdicate its functions of interference, but may guide the forces and laws of Nature," he replied.[123] Lyell was tied and devoted to Darwin by class, but he was truly wedded to Wallace in spirit. Despite Darwin's consternation, Wallace remained undeterred. Over the years he continued to develop a thoroughgoing teleological worldview that encompassed cosmological and biological realms.

Darwin must have wondered what had gotten into Wallace, but he failed to appreciate that the "child" born of the Ternate letter was a very different offspring from Darwin's. Despite the presumed similarity between the two hypotheses, the ideas expressed at the Linnean Society that summer evening of 1858 were really very different.[124] For one thing Wallace never used the term "selection" in his original formulation. For another (and more importantly) Wallace rejected Darwin's use of do-

mesticated animal breeding as a proof for the operations of natural selection. (More about this second point shortly.)

Of course, both Darwin and Wallace argued that their theories
were principles based upon a constantly changing environment along
with very small variations that affected individual survival and response
to environmental pressures resulting in differential death rates and
moreover that species held a tendency to form new perpetuating varieties. These resulted in adaptive progress for surviving species while at the
same time causing a branching indefinite divergence of new species. In
short, varieties would eventually over time convert into new species.

However, Darwin and Wallace each read Malthus differently. Darwin considered the food supply had to be on average constant with the
increase of population geometrically; Wallace on the other hand saw the
growth or depletion of a population due to available food and the ability of a given species to exploit it. In short, Darwin saw competition as
taking place between individuals while Wallace saw competition as taking place between populations; thus competition led to modifications
of a group under Darwin's view with competition leading to changes in
population size of several groups for Wallace. Darwin focused on individual struggles for existence while Wallace concentrated on population
growth as the powerful modifying force in nature. Wallace saw evolution taking place not in an individual but in a demographic context. Both
views had problems. Wallace failed to clearly distinguish varieties and
variations; Darwin's hypothesis was premised upon the inheritance of
acquired characteristics, a notion exploded years later by August Weismann's failed attempt to confirm the phenomenon in the experimental
removal of successive generations of rat tails. Members of the Linnean
Society hadn't noticed any of these differences in 1858.

Wallace, the founder of biogeography, knew that domestic animals
had a tendency to revert to their original stock if placed in a wild environment or else perish. But this would not work in reverse; in other
words, wild species variation cannot be deduced from domestic practices because the very state of "selection" and then of subsequent feeding

and protecting of the newly bred animals effectively shelters them from the effects naturally bearing upon their survival. Wallace came to profoundly disagree with Darwin over his breeding examples as a proof of natural selection; all they demonstrated was *un*natural selection. Wallace emphasized the principle of utility. Wallace always insisted that domestication introduced an artificial effect; once this human intervention is removed species either revert to their "original type" or become extinct.

Darwin replied that his domestication examples proved that "hereditary modification" was possible and that artificial selections show that small variations can accumulate to change the species' type.[125] Darwin always believed that a general theory of selection was possible; Wallace always believed Darwin's example of domestic breeders to be naively anthropomorphic.

Darwin nevertheless remained adamant; he insisted that "unconscious selection" produced "domestic races" that have been modified by breeders and horticulturalists for years and that history showed that domesticated breeds have changed dramatically through several thousand years. Indeed Darwin went so far as to claim that the line between unconscious selection and natural selection was difficult to discern. As Jean Gayon has convincingly argued, Darwin's domestication examples were not simply metaphorical pedagogical devices, they were essential to the theory itself.[126] In words that Wallace himself could have penned, Phillip Johnson noted years later that "The analogy to artificial selection is misleading. Plant and animal breeders employ intelligence and specialized knowledge to select breeding stock and to protect their charges from natural dangers. The point of Darwin's theory, however, was to establish that purposeless natural processes can substitute for intelligent design."[127]

Of course, Darwin never did explain precisely *how* and *when* microvariations would produce macro-speciation and still leave his chance or random modification theory inviolate, which was his central thesis. By comparison Wallace's application of the principle of utility is, if nothing else, more consistent with the principle itself. That is to say, Wallace's formulation was rooted in experience better suited to the kinds of massive—even global—kinds of macroevolutionary species change that made Darwin's theory the unique and controversial idea that it was.

In an amazingly perceptive article, Melinda B. Fagan has found that the differences in the two naturalists' theories were deeply integrated into their collecting goals, objectives, and daily field practices. While Darwin came to the *Beagle* under no particular financial constraint or expectation, Wallace was, at least in part, in this for the money. Thus Wallace tended to collect twice: once for the museum and collector trade, the other for his own scientific collection. Because numbers were of the essence, Wallace's dawn to dusk collecting routine was essential. This demanded very different collecting styles: Darwin principally collecting along the coast and spending nearly half his time on board, Wallace working much longer, harder, and more intensely often in remote regions deep in the interior. Fagan points out that "Wallace's theoretical and economic interests led him to collect whole series of specimens for particular species, from his first expeditions on the Rio Negro in the 1840s, to his hunt for Paradise birds in the Aru Islands over a decade later."[128] While individuals are not unimportant for Wallace, he instead "consistently emphasized groups of organization, while Darwin described many details of individual organisms. Also, Wallace clearly distinguished between groups of organisms, while Darwin was more ambiguous."[129] Wallace emphasized species represented by a "good series" of many individuals, thus he "used populations of specimens to represent species, not one or two individuals, as Darwin did."[130] Fagan concluded that Wallace's theory "was neither confused nor misguided. Nor does it posit an additional process occurring over and above selection on individual organisms. After describing selection on individual organisms (an unusual departure from his typical emphasis), Wallace shifts to species and varieties, the focus of most of his writing, which his routine practice led him to emphasize."[131] Given their distinctive purposes and *modus operandi*, then, the differences between Wallace's and Darwin's natural selection become understandable.

Regardless of the differences, in both cases common descent along with its requisite macroevolutionary change remains a given. But the differing views of Darwin and Wallace on the principle of utility would compel a much deeper rift. Wallace came to realize something that biochemist Michael Behe would note well over one hundred years later: "Common descent is true; yet the explanation of common descent—even the common descent of humans and chimps—although fascinating, is in a profound sense *trivial*. It says merely that commonalities were there from the start, present in a common ancestor. It does not even begin to explain where those commonalities came from, or how humans subsequently acquired remarkable differences. *Something that is nonrandom must account for the common descent of life*."[132]

What is that "something"? Here Wallace had given his answer in the conclusion to his review of Lyell's *Principles of Geology*: "Let us fearlessly admit that the mind of man (itself the living proof of a supreme mind) is able to trace, and to a considerable extent has traced, the laws by means of which the organic no less than the inorganic world has been developed. But let us not shut our eyes to the evidence that an Overruling Intelligence has watched over the action of those laws, so directing variations and so determining their accumulation, as finally to produce an organization sufficiently perfect to admit of, and even to aid in, the indefinite advancement of our mental and moral nature." It was an answer he would spend the rest of his life elaborating.

So what *really had* Wallace developed in terms of an evolutionary theory? The best approach is to first define Darwin's theory. It should be made clear from the outset that Darwin's evolutionary theory operated by three related propositions: 1) species were mutable; 2) evolution extends to account for virtually *all* biodiversity; and 3) the process of change was caused by natural selection and random variation.

It is most important to bear in mind that Darwinian evolution functions through variation, a wholly "random" process.[133] Blind variations (mutations, according to modern Darwinists) operating through natural selection effectively render William Paley's argument from design moot.

Giving over biological life to randomness and change wasn't especially directed at eliminating the role of a Creator or teleological purpose in nature, simply to make such considerations superfluous in light of a particular type of scientific inquiry called methodological naturalism, the notion that scientists *must* invoke *only* unintelligent material processes functioning via unbroken natural laws in nonteleological ways. But Wallace's suggestion of an "Overruling Intelligence" in the process of developing the human mind challenged Darwin's evolutionary framework, a framework that served not only to bolster a materialistic metaphysic but, in effect, proposed to become its operative manifesto. Indeed in the end, it supports the inescapable conclusion that Darwinian evolution far from being a scientific theory is "one long argument" in favor of an *a priori* metaphysic.[134]

Darwin's own words on the subject support this conclusion. "With respect to the theological view of the question. This is always painful to me. I am bewildered. I had no intention to write atheistically. But I own that I cannot see as plainly as others do, and as I should wish to do, evidence of design and beneficence on all sides of us. There seems to me too much misery in the world…. I am inclined to look at everything as resulting from designed laws, with the details, whether good or bad, left to the working out of what we may call chance."[135]

Whether, given his Plinian experiences, Paley ever *was* "conclusive," Darwin in his typically disingenuous way claimed, "The old argument from design in Nature, as given by Paley, which formerly seemed to me so conclusive, fails, now that the law of natural selection has been discovered…. There seems to be no more design in the variability of organic beings, and in the action of natural selection, than in the course which the wind blows."[136] Darwin was also very concerned to dispel any false impressions he may have left with regard to teleology in nature. "For brevity sake," he explained, "I sometimes speak of natural selection as an intelligent power; in the same way as astronomers speak of the attraction of gravity as ruling the movements of the planets, or as agriculturalists speak of man making domestic races by his power of selection. In the one case, as in the other, selection does nothing without variability, and this depends in some manner on the action of the surrounding circumstances on the organism. I have, also, often personified the word Nature; but I mean by nature only the aggregate action and product of many natural laws—and by laws only the ascertained sequence of events."[137]

The soft-spoken patriarch of Down House always tried to downplay the philosophical and religious aspects of his theory. Darwin wanted acceptance above all, and to achieve that he was willing to engage in any number of strategies. One of the most obvious was to insert into the second edition of his *Origin* some language to placate the clergy over the implication of his work. In the first edition, Darwin simply closed his book with, "There is grandeur in this view of life, with its several powers, having been originally breathed into a few forms or one...." But in the very next edition, published on January 7, 1860 (only about six weeks after the first), Darwin added, "There is grandeur in this view of life, with its several powers, having been originally breathed *by the Creator* [emphasis added] into a few forms or one...."

Later, in a letter to Joseph Hooker on March 29, 1863, Darwin claimed his regret that he had "truckled to public opinion & used the Pentateuchal term of creation, by which I really meant 'appeared' by some wholly unknown process. It is mere rubbish thinking, at present, of origin of life; one might as well think of origin of matter."[138] If Darwin regretted all his "truckling to public opinion" so much, why did he never remove the term from the four subsequent editions of his *Origin*? The only conceivable answer is that Darwin preferred the public relations advantage such "truckling" offered.

None of this, of course, suggests that Darwin was open to any kind of teleology in his brand of evolution. He had long dismissed that possibility, and the evidence found in his own private notebooks, largely compiled upon his return from the *Beagle* voyage in 1836, is replete with favorable—at times even enthusiastic—references to the skeptic David Hume and atheist/positivist Auguste Comte.[139] No wonder that Cyril Darlington, an otherwise sympathetic Darwinian, called him (com-

pared to Wallace, Lyell, and Hooker) a "slippery" character whose verac-
ity and intellectual integrity was not to be trusted.[140]

In this context Wallace's "murder" becomes immediately apparent
when we see precisely what he was formulating in contradistinction from
the materialistic methodological naturalism of Darwinism. Wallace instead proposed a theory of common descent based upon natural selection strictly bounded by the principle of utility within a larger teleological and theistic framework. It was, in fact, largely a revision of his earlier
essay on "The Origin of Human Races and the Antiquity of Man." But,
as Martin Fichman has observed, Wallace constructed a theistic evolutionary model that made natural selection subservient to much higher
teleological directive powers.[141] *Intelligent* evolution was born with the
April issue of the *Quarterly Review*; the immediate catalyst was, quite
appropriately, a work by Charles Lyell who had suggested an evolution
imbued with intelligence in his own work.

Darwin could complain about Wallace's defection, but the renegade
captain could remind the admiral that, after all, it was the admiral who
had woven the principle of utility (the principle that suggested the "murder" in the first place) into the very fabric of his natural selection theory:
"I think it would be a most extraordinary fact if no variation ever had occurred useful to each being's own welfare," Darwin insisted in his *Origin*,
"in the same way as so many variations have occurred useful to man. But
if variation useful to any organic beings do occur, assuredly individuals thus characterized will have the best chance of being preserved in
the struggle for life; and from the strong principle of inheritance they
will tend to produce offspring similarly characterized. This principle of
preservation, I have called, for the sake of brevity, Natural Selection."[142]
From this standpoint utility *was* natural selection.

Still, Wallace's insertion of an intelligent cause and agency even if
located behind or even through natural laws was treason for Darwin's
recalcitrant materialism. Not because it was *unscientific* per se; there
was nothing "unscientific" about limiting the principle of utility. Darwin himself was increasingly relying upon subsidiary theories of sexual

selection and pangenesis to shore up his theory (both of which Wallace thought were seriously flawed). It was treason because Wallace's suggestion ran counter the philosophical assumptions of materialism and methodological naturalism that were inherent in Darwin's theory. Darwin's evolutionary theory wasn't just a story of common descent; it was a vindication for the blind forces of materialism.

The father of Darwinism did not bring into this world some innocent offspring of a dispassionate search for scientific truth. As we have seen, Darwin was seduced by Plinian freethinkers into birthing a child of a particular kind; a soulless child that saw everything, including man himself, as a product of soulless processes. Wallace's "Overruling Intelligence" would have slain such a pernicious demon.

Would have but didn't. Why? The answer is complex and by no means amenable to a simple answer. However, a big part lies in the formation of the X Club on November 3, 1864. Meeting at six or so each first Thursday of the month, members of Darwin's inner circle met for dinner at Saint George's Hotel at Abermarle Street.[143] The roster included Thomas Henry Huxley and Joseph Hooker (Darwin's most intimate confidantes); John Tyndall (a close friend of Huxley), George Busk (close friend of Hooker and Linnean Society secretary who read the Darwin-Wallace papers at the unveiling of natural selection), Edward Frankland (a chemist and friend of Tyndall), Herbert Spencer (philosopher and friend of Huxley and Wallace), Thomas Hirst (mathematician and friend of Tyndall who was converted to transmutation with reading Robert Chambers's *Vestiges*), and John Lubbock (the well-connected son of Sir John Lubbock, 3rd Baronet, a neighbor of Darwin's and a frequent visitor to Down House). Conspicuously absent were Charles Lyell (much older than the rest and never enthusiastic for evolution) and Alfred Russel Wallace (by temperament indisposed to such gatherings and after 1869 anathematized by this tightly knit group).

Darwin was not one to directly engage his enemies. Rather, he sent his loyal captains to do his bidding. The two principals were Huxley and Hooker. Huxley loved the fray, and spoke to Darwin about "sharpening my claws and beak in readiness."[144] At the lecture podium he was a rough-and-tumble street brawler. Known for speaking extemporaneously and making quickly but adroitly sketched illustrations of his points, he could also heap on abuse with his rapier wit.

Hooker was more tactful and discrete in dealing with the opposition. He was always willing to lend his support, but his quieter style sometime prompted Darwin to wonder, "I feared that you were weary of the subject."[145] But Hooker's less in-your-face approach had advantages. When he was unable to get the editor of *Gardener's Chronicle* to run a notice on a recent work by Darwin, he simply wrote it himself attempting to mimic the editor's style.[146]

By the end of the 1860s the X Club had won Darwin's battle. The speed with which the professional public and even Huxley's "educated mob" was browbeaten into acquiescence was nothing short of remarkable. The prize was "science as synecdoche for Darwinism."[147] Bandying about the term "science" as synonymous with Darwinian materialism, Huxley spoke more truth than perhaps he intended when he declared, "The English nation will not take science from above, so it must get it from below. We, the doctors, who know what is good for it, if we cannot get it to take pills, must administer our remedies par derriere...."[148] With unabashed hubris Huxley readied his Darwinian syringe and not too politely asked the public to bend over.

Darwin was not satisfied with winning mother England. He also encouraged the spread of the Darwinian gospel to the Continent. In 1862 the French atheist Clémence Royer released a translation of *Origin of Species* to the applause of Paul Broca's Society of Anthropology. In Germany Ernst Haeckel, a deterministic anti-Christian monist, played fast and loose with embryo drawings to support his now-discredited evolutionary recapitulation theory. Haeckel also championed Darwin's cause in Germany. If Huxley was Darwin's "Bulldog" then Haeckel was surely his "Dachshund." In 1864 Haeckel happily wrote to Darwin that the "best" of Germany's youth were committed to Darwinism.[149] His *General Morphology of Organisms* (1866) was one of Germany's first biology texts written from a Darwinian perspective, and (like Huxley) he gave a series of popular lectures on evolution, which were compiled and published as *The Natural History of Creation* (1868).[150]

By the time Wallace formally broke with Darwin, Darwinism had been victorious. Wallace's marginalization from Darwin's circle was certain. He was, in Ross A. Slotten's words, "the heretic in Darwin's court." Wallace knew this. In a letter dated April 28, 1869, responding to Darwin's dismay over his *Quarterly Review* piece, Wallace pulled no punches and wrote out an even more thorough explanation of his views:

> It seems to me that if we once admit the necessity of *any* action beyond "natural selection" in developing man, we have no reason whatever for confining that agency to his brain. On the mere doctrine of chances it seems to me in the highest degree improbable that so many points of structure, all tending to favour his mental development, should concur in man alone of all animals. If the erect posture, the freedom of the anterior limbs from purposes of locomotion, the powerful and opposable thumb, the naked skin, the great symmetry of form, the perfect organs of speech, and, in his mental faculties, calculation of numbers, ideas of symmetry, of justice, of abstract reasoning, of the infinite, of a future state, and many others, cannot be shown to be each and all *useful* to man [on the principle of utility] in the very lowest state of civilization—how are we to explain their co-existence in him alone of the whole series of organized being? Years ago I saw in London a bushman boy and girl, and the girl played very nicely on the piano. Blind Tom, the half-idiot negro slave, had a "musical ear" or brain, superior, perhaps, to that of the best living musicians. Unless Darwin can show me how this latent musical faculty in the lowest races can have been developed through *survival* of the fittest, can have been of *use* to the individual or the race, so as to cause those who possess it in a fractionally greater degree than others to win in the struggle for life, I must believe that some other power (than natural selection) caused that development. It seems to me that the *onus probandi* will lie with those who maintain that man, body and mind, could have been developed from a quadrumanous animal by "natural selection."[151]

Wallace reflecting on that letter in 1908 thought it the best and most succinct statement on his position than anything he had up to then published. Indeed he is correct. Whether living among Uaupés River natives or Dyak headhunters, Wallace remembered their capacities for reason, music, language, altruism, and hosts of other uniquely human attributes as being at one with his own. In contrast, Darwin saw the "lowest" of natives as reflective of primordial social instincts "developed by nearly the same steps" as the "lower animals."[152] This is not to suggest that Darwin was not keenly aware of the differences of degree in *Homo sapiens* and animals; Darwin was, after all, a proponent of the unity of humankind as a species and a vocal opponent of slavery. In fact Adrian Desmond and James Moore have recently made a case for the "*humanitarian* roots" of Darwin's evolutionary theory in *Darwin's Sacred Cause*.[153]

Yet by their own account the difference in approach to the so-called "savage" races between Wallace and Darwin is unmistakable and leaves Darwin hardly a racial egalitarian. Desmond and Moore write, "... like many, Darwin equated 'savagery' in its 'utter licentiousness' and 'unnatural crimes' with the values of his own under-class (two groups the socialist Wallace held in high regard). But by lowering 'savage' morality and raising ape capabilities, Darwin made the continuum towards civilization seem more feasible. It was humanitarianism that Darwin took pride in.... Yet the incongruity of his class holding this ethic sacrosanct while disparaging the 'lower' races (even as colonists displaced or exterminated them) is impossible to comprehend by twenty-first century standards."[154] Indeed, between Wallace and Darwin it is the former who appears more modern and in accord with current sensibilities. One is left wondering how "sacred" Darwin's cause really was; Desmond and Moore themselves are forced to admit that, "while slavery demanded one's active participation, racial genocide was now normalized by natural selection and rationalized as *nature's* way of producing 'superior' races. Darwin ended up calibrating human 'rank' no differently from the rest of his society."[155] Wallace's and Darwin's different attitudes were symptomatic of different worldviews. For Wallace, humanness was something apart from the ordinary biological world; for Darwin this simply was not the case.

Wallace would state his case more publicly in an essay, "The Limits of Natural Selection as Applied to Man", published in 1870 as part of an anthology of his works titled *Contributions to the Theory of Natural Selection. A Series of Essays*. These ten works give a thorough representation of Wallace's work up until that time. It includes the famous Ternate letter (see Appendix A) and besides his work on human development, comprises essays related to ornithology, Lepidoptera, and animal mimicry.

With Darwinism secure, the Down House patriarch finally tackled the application of his theory to humans in *The Descent of Man* (1871). No doubt with Wallace in mind Darwin wrote, "Spiritual powers cannot be compared or classed by the naturalist; but he may endeavor to show, as I have done, that the mental faculties of man and the lower animals do not differ in kind, although immensely in degree."[156] Wallace would never agree.

Despite their disagreements, Darwin and Wallace maintained gentlemanly correspondence for the remainder of their lives. For Wallace's part, he commented, "It is really quite pathetic how much he felt difference of opinion from his friends."[157] Darwin *did*, however, always feel a bond and to some extent an obligation to Wallace. After all, it *was* this unknown specimen collector in the wilds of the Ternate and Gilolo islands that finally prompted him to action, and had Wallace been as possessive of *"my theory"* as Darwin, things could easily have taken a nasty turn. But they didn't and Darwin was always thankful for that.

Soon Wallace had something to thank Darwin for. Arabella Buckley, secretary to Charles Lyell, an intelligent young woman who wrote reviews and children's books on nature, came to know Wallace. Her mother's pursuit of spiritualism no doubt gave Miss Buckley and Lyell's colleague a great deal to talk about. How Arabella came to know of Wallace's financial difficulties is unclear, but in 1879 she wrote (unbeknownst to Wallace) a letter to Darwin asking that the famed naturalist assist in obtaining a government pension for her friend. Wallace always had a steady income from his writing, but he was a poor investor and had spent a large sum of his personal savings litigating against an insane zealot attempting to prove the earth flat. Darwin didn't hold out much hope but offered to ask Hooker what might be done. There the matter rested.

Then in 1880 Wallace published *Island Life*. Issued as a sequel to his *Geographical Distribution of Animals* (1876), it was a sensation. Darwin thought it was Wallace's best work. Wallace dedicated the book to Hooker, and Hooker agreed with Darwin on its merit. This formed the stimulus for Darwin to revive the idea of a pension a second time. This round worked. Huxley, who always thought Wallace deserved the honor and recognition of England for all he had done, led the charge. Arabella Buckley drew together a *curriculum vita* of his accomplishments. Three stood out: the sheer size of his Malay Archipelago collections, his independent discovery of natural selection, and his application of natural selection to the geographical distribution of animals (biogeography). Darwin sent a personal note to Prime Minister Gladstone and backed it up with well placed letters to everyone he could think of who might aid in the pension process. Darwin received a reply on January 7, 1881:

> Dear Mr. Darwin
>   I had in some degree considered the subject of your note and the memorial upon it [sic] arrival and I lose no time in apprising you that although the Fund is moderate, and is at present poor, I shall recommend Mr. Wallace for a pension of £200 a year. I remain
> Faithfully yours
> W. E. Gladstone[158]

It was certainly gratifying for Darwin and welcomed relief for Wallace. In 2010 dollars this amounts to an annual pension of $23,400.00 (see Measuring Worth), well above the average American wage earner's annual salary during the period of $832. The pension lifted a real burden from the Wallace household. Wallace would, of course, continue to write but it meant that he was free from the ever-present necessity of writing to produce an immediate income.

Darwin's last gesture toward Wallace was a very kind one, one he could have easily ignored. When Darwin got the final approval from Gladstone he was not a well man. He would lose his loveable but ne'er-do-well brother Erasmus in August, and in October his "worm book" (*The Formation of Vegetable Mould Through the Action of Worms*) was released. Tired, winded, and with a weak pulse, Darwin continued to work as far as his condition would allow. "I am fairly well," he wrote to Wallace in January 2, 1881, "but always feel half dead with fatigue."[159] He would struggle on another year. On the afternoon of April 19, 1882, Darwin died. There were rumors of a deathbed conversion from Darwin's stated agnosticism, but they were likely more myth than fact.[160] Wallace served as pallbearer at the funeral held at Westminster Abbey on April 26; thus closed the relationship between Alfred Russel Wallace and Charles Darwin.

Their lives had both been complex and, at times, full of controversy. So too their friendship was a complicated one. On one level Darwin would never forgive Wallace for his mutinous defection. It wasn't a question of different scientific opinions; it was a question of different worldviews. What precisely *was* Darwins' worldview?

As we have seen, the radical deism of his grandfather Erasmus festered into the quiet atheism in his father Robert, and as a boy the Unitarian instruction of young Charles devolved to his sisters. Introduced to radical freethinkers as a teenager in the Plinian Society during his abortive attempt at pursuing a medical career at the University of Edinburgh, we find him taking almost naturally to the skepticism of David Hume and the positivism of Auguste Comte.[161] No wonder Janet Browne admits that "Darwin was profoundly conditioned to become the author of a doctrine inimical to religion."[162] Darwin claims to have started out as a theist when writing *Origin of Species*, but then asks rhetorically, "can the mind of man, which has, as I fully believe, been developed from a mind as low as that possessed by the lowest animals, be trusted when it draws such grand conclusions?" He concluded, "I for one must be content to remain an Agnostic."[163]

But can we really leave it at that? In the end, can we conclude that Darwin, in a hopeless theological muddle, simply settled on uncertainty in this question? Some who read his *Origin* would have accepted perhaps a different designation. Adam Sedgwick blasted Darwin's theory. Accepting Darwin's evolutionary ideas threatened, according to Sedgwick, to "sink the human race into a lower grade of degradation than any into which it has fallen since its written records tell us of its history."[164] Charles Hodge, principal of Princeton Theological Seminary and America's leading Calvinist, agreed. Whatever Darwin's personal religious faith may or may not be, he insisted, Darwinism *is* atheism.[165] So what are we to make of Darwin? Was he atheist or agnostic?

On balance, the historical evidence suggests that Darwin's religious views always tended toward some form of theistic nihilism. Darwin was always careful to keep any teleological implications out of his theory. It is clear that when Darwin viewed nature, God was not there. In fact, for Darwin, man was mere animal, different in degree certainly but not in kind. As for the complex emotions often associated with reverence for God, Darwin saw parallels in the "deep love of a dog for his master" and "of a monkey to his beloved keeper."[166] "The idea of a universal and beneficent Creator," he insisted, "does not seem to arise in the mind of man until he has been elevated by long-continued culture." In short, God is the invention of man *not* man the creation of God. All this tends toward atheism. But to view Darwin simply as an atheist and leave it at that seems too simplistic. After all, he claimed to be an agnostic. Why not take his word for it?

The problem with simply calling Darwin an agnostic is that agnosticism means many things. Thomas Henry Huxley, in fact, coined the word to distance himself from charges of materialism and even atheism. But it became a failed strategy as agnosticism soon came to have a wide range of connotations in public discourse and common parlance. Even Lenin noticed the miscarriage stating that "in Huxley agnosticism serves as a fig leaf for [his] materialism." Indeed by the end of the nineteenth century agnosticism had come to mean different things to different people. Many simply regarded agnosticism as a kind of uncertainty about God's existence; hypothetically any agnostic might be swayed into belief by reason and argument. At first blush one is inclined to associate Darwin with this brand of agnosticism. Darwin, after all, was always a minimalist in his negation of God. However, he never felt a direct attack was necessary because he, like Huxley, believed that all talk of God and deity was beyond human understanding. Darwin adhered not to a weak form of agnosticism that says merely, "I don't know if there's a God because I've not seen sufficient evidence for Him;" his was a much stronger form of agnosticism that argued God was *unknowable*–all God-talk was ultimately, for Darwin, nonsense. It is this epistemological certainty that makes this a *strong* version of agnosticism. So here's the problem: simply calling Darwin an agnostic is not specific enough because it leaves the two forms (the strong and the weak) ambiguous.

Well known historian of science Maurice Mandelbaum understood this. In an interesting analysis of Darwin's religious views, he noted, "In the end his [Darwin's] Agnosticism was not one brought about by an equal balance of arguments too abstruse for the human mind; it was an Agnosticism based on an incapacity to deny what there was no good reason for affirming. Thus, those who, at the time, regarded Agnosticism as merely an undogmatic form of atheism would, in my opinion, be correct in so characterizing Darwin's own personal opinion."[167] Darwin as "undogmatic atheist" came as close to the truth as anyone had been able to come in the century since *Origin* appeared.

But perhaps another designation would be even more precise or at least equally useful in this regard. Scottish theologian Robert Flint offered a term of his own that comports well with Darwin's position. He wrote:

> The atheist is not necessarily a man who says "There is no God." What is called positive or dogmatic atheism, so far from being the only kind of atheism, is the rarest of all kinds. It has often been questioned whether there is any such thing. But every man is an atheist who does not believe that there is a God, although his want of belief may not be

rested on any allegation of positive knowledge that there is no God, but simply on one of want of knowledge that there is a God. If a man have failed to find any good reason for believing that there is a God, it is perfectly natural and rational that he should not believe that there is a God; and if so, he is an atheist, although he assume no superhuman knowledge, but merely the ordinary human power of judging evidence. If he go farther, and, after an investigation into the nature and reach of human knowledge, ending in the conclusion that the existence of God is incapable of proof, cease to believe in it on the ground that he cannot know it to be true, he is an agnostic and also an atheist, an agnostic-atheist—an atheist because an agnostic. There are unquestionably many such atheists. Agnosticism is among the commonest apologies for atheism. While, then, it is erroneous to identify agnosticism and atheism, it is equally erroneous so to separate them as if the one were exclusive of the other: that they are combined is an unquestionable fact.[168]

Flint's important study of *Agnosticism* offers an insightful and useful designation in the term *agnostic atheist*. Nick Spencer's recent article in *The Guardian* (interestingly cited approvingly on Richard Dawkins's blog May 21, 2009) noted a problem with the overly simplistic use of the term agnostic. "Attitudes are fine," he suggests, "but they need to be about something. Adjectives need nouns. If Huxley was indeed an agnostic, he was an agnostic atheist, tending away from the divine but unwilling (so he claimed) to be too dogmatic about it." And so too with Darwin.

Perhaps more importantly Darwin*ism* is suffused with agnostic atheism. Edward Larson is right in concluding that, "For Darwin, differential death rates caused by purely natural factors created new species. God was superfluous to the process."[169] Darwin never argued *against* God in any of his works, including *Descent of Man*, only against the necessity of God. This minimalist formulation is powerful in its dismissiveness of deity and thus forms an essential (though not necessarily sufficient) foundational premise for secularism. It was—and is—atheism but always of a distinctly undogmatic stripe. When the liberal Victorian clergy rushed to support *Origin*, Darwin was quick to respond. The Reverend Charles Kingsley approved of a theistic brand of Darwinism, and sure enough it soon found its way into the very next edition of *Origin* in January of 1860 (and every subsequent edition thereafter) as having the approbation of a "celebrated author and divine." When Harvard botanist Asa Gray supported his own theistic version of *Origin*, Darwin compiled his warmly supportive reviews and published them as *Natural Selection Not Inconsistent with Natural Theology. A Free Examination of Darwin's Treatise on the Origin of Species, and of Its American Reviewers* in 1861. Publication expenses were completely borne by Darwin. As Benjamin Wiker points out, it's not that Darwin actually agreed with Gray; his private correspondence is replete with his polite objections to Gray's theistic additions. Nevertheless, "he had no qualms about using Gray's argument if it would smooth the way for acceptance of his theory. Once the theory was accepted," Wiker adds, "the theistic patina would be ground away by the hard, anti-theistic core of the argument."[170] The point is it would be wrong to interpret Darwin's willing inclusion of Kingsley's religious support in *Origin* or his eager approval of Gray's theistic reviews of his work as evidence of his matching belief; Darwin was always more than willing to set his hard agnosticism aside in the interest of promoting his pet theory.

So what are we to make of Darwin's religious beliefs? There are five possibilities:

1. Darwin was a *religious believer*. This is hardly supportable by any historical evidence whatsoever.

2. Darwin was an *agnostic*. This is true as far as it goes, but the term itself is too vague and diverse in meaning to be of much use and, in fact, may leave seriously misleading impressions.

3. Darwin was an *atheist*. This is also true insofar as his theory tended to support atheism but probably goes too far in relation to Darwin himself for it implies a dogmatism ill-suited to his subtler and more pragmatic nature. For all of Richard Dawkins's effusions on behalf of the Down House patriarch,

Darwin would likely have found Dawkins's approach crude and unappealing if not downright appalling.

4. Darwin was an *undogmatic atheist*. This apt phrase suggested by Mandelbaum is descriptive of Darwin's belief and approach but must be reconciled with his own claims to being "an Agnostic."

5. Darwin was an *agnostic atheist*. This comes closest to encompassing the range and character of his beliefs and it comports to his theory as well.

So, in the end, it is fairly easy to accept either Darwin as *undogmatic atheist* or *agnostic atheist*. The dual attribution of "atheism" shows the common ties that bind. But the wishful pleadings of Darwinian evolutionists like Karl Giberson and others that Darwin was a "sincere religious believer" whose eventual conversion to a more hardened agnosticism was late in life and reluctant are utterly without historical merit.[171] As noted earlier, Darwin's notebooks demonstrate quite clearly his religious skepticism and materialistic propensies as early as age 29, ideas he had been introduced to as early as age 17 as a Plinian. The Plinian Society was telling for Darwin. Despite his casual dismissal of them in his *Autobiography*, Darwin was exposed to some of the most radical free-thinking of day at those meetings. Darwin was always careful to conceal this fact because its revelation would have made plain the philosophical template through which he would make all his observations while voyaging on *The Beagle*. In short, the metaphysic *preceded* the science.

Is Darwinian evolution compatible with theism? It surely was never intended to be and *certainly* never intended to be compatible with Christianity, though Darwin was more than willing to enlist religious allies on its behalf. Darwin's materialism would sharpen into the undogmatic atheism or agnostic atheism described above, but materialism was the template upon which he developed his evolutionary theory to be sure. Whether Darwin was a full-blown materialist or, as Neal Gillespie believes, a positivist influenced by the ideas of Comte is to argue philosophical details that largely amount to the same thing, but Darwin was most surely not a weak or soft agnostic who abandoned his faith slowly and reluctantly.

With Darwin in context it can be easily seen how distant Wallace had become from his senior's ideas. For all of Darwin's kindhearted support for Wallace's pension, the co-discoverer of natural selection knew his beliefs marginalized him from the new seat of scientific power and authority. For his part, Wallace always felt that he was being a more thorough explicator of Darwinian principles. He never thought much of Darwin's sexual selection and he rejected the notion of pangenesis. Wallace sided with Weismann on inheritance against Darwin's inheritance of acquired characteristics. Although Weismann failed to uncover the precise mechanisms of inheritance, he was the first to correctly outline the genetic transmission process. Slotten is correct in stating that "Wallace was among the first to recognize Weismann's genius and actively promote his ideas."[172] It would take Gregor Mendel, another opponent of Darwinian evolution, to elucidate the exact processes of genetic inheritance.[173]

After Darwin's death, Wallace directed most of his attentions to further expanding on his brand of Darwinism. In 1889 he published his fullest explication yet: *Darwinism: An Exposition of the Theory of Natural Selection with Some of Its Applications.* In his chapter XV he gives his complete views on "Darwinism Applied to Man." (See Appendix B for the complete excerpt of pages 473–478 in the original book.)

After outlining the many common features of *Homo sapiens* to other mammals (homological vertebrate and muscular structures, etc.), Wallace admits this is all strongly suggestive of descent from some common primordial primate. Here he is in agreement with Darwin. However, he goes on to argue that humans were the result of unique and special forces operating beyond the capacity of natural selection. The moral and intellectual capabilities of humans are unique, he argued, and are inexplicable by the principle of utility. That is to say, the moral and higher intellectual attributes of mankind do not convey any real survival advantage over their natural competitors. What survival advantage, he asks, do mathematical, musical, or artistic abilities afford? What advantage is gained by abstract reasoning or moral sensibilities? After presenting evidence that none of these uniquely human attributes could have been produced by natural selection, Wallace concludes that these can only be accounted for by some "spiritual influx" to which "the world of matter is altogether subordinate." In short, the mind of man was inexplicable by mere survival of the fittest. Moreover, this spiritual influx was discernible in three stages of the organic world: first, in the origin of life; second, in development of consciousness, "the fundamental distinction between the animal and vegetable kingdoms"; and finally in the existence of humankind, a class different from all animal existence that is unique and unbridgeable. Wallace absolutely rejected the notion that *Homo sapiens* were the product of blind or random processes, calling it a "hopeless and soul-deadening belief" without scientific evidence or merit.

Wallace had now gone beyond man to include the origin of life and sentience in animals as clear entry points for design and purpose. While Wallace may have thought that none of this opposed Darwinian theory, others disagreed. There was talk of *Wallaceism*. Darwinian critic and author of the utopian satire *Erewhon* (1872) Samuel Butler and Dutch zoologist A. A. W. Hubrecht both used the term. But the term had been floating about even before Wallace's chapter in *Darwinism*. During Wallace's highly successful tour of the United States from the fall of 1886 through the spring of 1887, the *Boston Evening Transcript* reported November 2, 1886, on Wallace's Lowell Lecture, calling "the first Darwinian" a master "of condensed statement—as clear and simple as compact—a most beautiful specimen of scientific work." Noting Wallace's position on the unique status of man versus the lower animals, it concluded that this was as lucid a presentation of "Wallaceism" as one could hope to hear.[174]

Nevertheless, Wallace refused the designation and even demanded an apology from Hubrecht. Wallace just refused to see that his theory was no longer Darwinian; Wallace had now become the champion of *intelligent evolution*, an evolutionary model intrinsically based upon intelligent design. Even his close friend Herbert Spencer tried to tell him. Upon receiving a copy of *Darwinism*, Spencer warned, "I regret that you have used the title 'Darwinism,' for notwithstanding your qualification of its meaning you will, by using it, tend greatly to confirm the erroneous conception almost universally current."[175]

As it was, Wallace's stubborn insistence upon equating his evolutionary theory with "Darwinism," was more obfuscating than elucidating. By doing so he consigned himself to the obscurity that the Darwinian banner would surely hold for him. It was exacerbated when George John Romanes, who sought to assume the mantle of leadership following Darwin's death, accused Wallace, quite misleadingly, of "ultra-Darwinism" for his strict selectionist views.[176]

The important point here is how this played out *in the application of natural selection to biological phenomena*. Wallace's "selectionism" was not really more "ultra," it was more sharply focused and specifically applied according to Darwin's own principles. Guided by the principle of utility, Wallace's application of natural selection was self-sufficient to explain *most* but—*and this is critical to appreciate*—*not* all (the three important exceptions previously noted) of the biological world. Darwin added pangenesis, but it was natural selection that remained central to his theory. Because Darwin was hidebound to methodological naturalism he, in effect, *had* to make natural selection do far more work than did Wallace. David Quammen is quite right when he says of Darwinian evolution, that the purposeless, impersonal and blind process of "natural selection isn't the sole mechanism of evolutionary change. But it's the primary mechanism. It's the lathe and the chisel that shape adaptation. It's the central concept of Darwinism, whatever else Darwinism might be taken to include. It's the starting point for understanding how evolution works."[177]

Nothing this strong could be said of Wallace's concept of evolution. For Wallace, natural selection was *limited* and *constrained* by profound teleological forces and factors. Thus, Darwin applied natural selection more indiscriminately to virtually *all* aspects of nature—e.g. man—

whereas Wallace *limited* and *targeted* its application. "Ultra-Darwinism," a phrase Romanes coined in the heat of argument, misleadingly implies that Wallace's *application* of natural selection was more "ultra" when, in fact, just the opposite is true. The correct phrase should be more "self-sufficient and specifically applied" *not* more "ultra." There is a profound difference. Unfortunately Wallace himself encouraged the conflation. "I believe," Wallace wrote, "that I have extended and strengthened it [natural selection]. The principle of 'utility,' which is one of its chief foundation-stones, I have always advocated unreservedly; while in extending the principle to almost every kind and degree of coloration [a reference to Darwin's sexual selection theory], and in maintaining the power of natural selection to increase the infertility of hybrid unions [a reference to Darwin's pangenesis], I have considerably extended its range. Hence it is that some of my critics [especially Romanes] declare that I am more Darwinian than Darwin himself, and in this, I admit, they are not far wrong."[178]

The essential problem is that this divides Darwinism across a false boundary. The question that animated Darwinism never was the extent to which natural selection could explain biological life and evolution but the degree to which a unified theory of evolution could be presented wholly resting upon naturalistic principles. Because Wallace tended to equate Darwinism with natural selection itself, he remained adamant in his loyalty to Darwinism, feeling that his was a more purist defense of the principle itself rather than sullying it with subsidiary notions of sexual selection and pangenesis.

What he seems to have not fully appreciated was the degree to which Darwin and his fellow captains were wedded to methodological naturalism. It has been suggested by some that Wallace's spiritualism "caused" his break with Darwin; more accurate is the fact that Wallace's exploration of spiritualism, which he always claimed he did from a thoroughly analytical and scientific basis, permitted him a less constrained view of science.

The interesting point is that both camps saw the weaknesses of natural selection as an all-explanatory mechanism. Darwin was forced to call upon subsidiary theories in its defense; Wallace simply discerned its limits and called upon a teleological argument to offer a more coherent view of nature. Historian Martin Fichman has perhaps put it best: "Theism completed Wallace's evolutionary worldview. He saw theism, in terms of intelligent design, as providing an account of the emergence of those human traits he deemed inexplicable by natural selection and necessary for the possibility of future human progress. Wallace came to regard intelligent design as guiding certain aspects of the development of the nonhuman organic world as well."[179] In this sense, Wallace was surely no "ultra-Darwinist." His detractors made this charge precisely because they refused to count his theistic additions as explanations, and this not based upon any incontrovertible evidence but upon their *a priori* commitment to methodological naturalism.

9.

INTELLIGENT EVOLUTION

MAN'S PLACE IN THE UNIVERSE AND THE WORLD OF LIFE: A GRAND SYNTHESIS

WALLACE APPROACHED HIS CONCEPT OF INTELLIGENT EVOLUTION from two standpoints: cosmologically and biologically. While he had given considerable thought to the latter aspects of his theory, he had said little about how a teleological world infused with intelligent design might be applied on a cosmic scale. In 1903 he addressed this question in *Man's Place in the Universe*. Here Wallace argued against the mis-named "Copernican Principle," that given the immensity and age of the universe there is nothing special about the earth or the life upon it. We may, in fact, not be alone in the universe; the likelihood of many other intelligent life forms suggests man's insignificance. (This is certainly *not* an argument that Copernicus himself ever made.)

In examining the latest astronomical data, Wallace disagreed. He found our solar system especially well positioned for the emergence of life. He then spent considerable time delineating and amplifying what he considered the five broad essentials for organic life: 1) precise tolerances of temperature; 2) sufficient solar light and heat; 3) abundant water supply; 4) a sufficiently dense atmosphere conducive to life; and 5) a planet with balanced alterations of day and night.[180] Once one factors in many "sub-conditions" necessary to support organic life, Wallace suggested that as many as fifty may be required. Taken altogether "the chances against the simultaneous occurrence of the whole fifty would be a million raised to the eighth power ($1,000,000,000^8$), or a million multiplied by a million eight times successively to 1. These figures are suggested merely to give some indication to the general reader of the way in which the chances against any event happening more than once mount up to unimaginable numbers when the event is a highly complex one [like the origin of life]."[181] Wallace proposed that the universe in which we live was designed for the development of humanity. To those who suggested that perhaps even under some theistic view humans were only one of many intelligent life forms, he said:

> But to those who believe that the universe is the product of mind, that it shows proofs of design, and that man is the designed outcome of it, and yet who urge that other worlds in unknown numbers have also been designed to produce man, and have actually produced him—to these I reply, that such a view assumes a knowledge of the Creator's purpose and mode of action which we do not possess; that we have no guide to His purposes but the facts we actually know; that we *do* know that here, on our earth, man is the culmination of one line of evolution, not of many, and that the presumption, therefore, is, that no line of evolution in other worlds under other conditions could produce him.[182]

Wallace had produced half of his argument. He knew to be complete he needed to apply his theory of intelligent evolution to *The World of Life*. It was issued from the London publishing house of Chapman and Hall on December 2, 1910.[183] If intelligent evolution was born with his essay review of Lyell's work, it came of age with *The World of Life*. Here in one cover Wallace weaved together the tapestry that formed the fabric of his life's thought and work. It may be justly said to represent the epitome not only of his evolutionary theory but also of his cosmology. When taken together these integrate into an overarching epistemological and ontological framework with moral and ethical implications for the sanctity of life and its proper development. A brief review of its chapters will outline the main features of this most important of Wallace's sizeable bibliography.

In many ways, the title says it all: *The World of Life: A Manifestation of Creative Power, Directive Mind and Ultimate Purpose*. Wallace sets forth his thesis from the outset: arguing from specific examples in the animal kingdom—the bird's feather, metamorphosis in insects, and other "marvelous transformations of the higher insects,"—that "they necessarily imply, first, a Creative Power, which so constituted matter as to render these marvels possible; next a directive Mind which is demanded at every step of what we term growth, and often look upon as so simple and natural a process as to require no explanation at all; and, lastly, an ultimate Purpose, in the very existence of the whole vast life-world in all its long course of evolution throughout the eons of geologic time."[184]

In chapter one Wallace sets the tone for his book with a biting critique of Haeckel's monism and Huxley's materialism, both of which he finds vague and unsatisfactory. Chapters two through ten give a thorough overview of evolution and evidence that Wallace believes supports it; his discussion of plant and animal distribution as well as various applications of natural selection and adaptation are particularly interesting. In chapter eleven he insists that organic life cannot be the result of "self-acting agencies" but must come about from some type of "mind-action."[185] In the next chapter Wallace asserts that "Mind" and "Purpose" lie beyond natural phenomena.

In the chapter "Birds and Insects: As Proofs of an Organising and Directive Life-Principle" Wallace makes an explicit case for design in various aspects of nature. He singles out birds and insects as demonstrable examples of "an organizing and directive life principle." Wallace notes the intelligence of many birds, rivaling that of numerous mammalian counterparts, but it is the bird's feathers that capture his greatest attention. The bird's feather and wing demonstrate, for Wallace, "a *preconceived design* [emphasis added]."[186] Reproducing an image of the intricate make-up of the feather with its detailed interlocking hook-and-eye mechanisms of the barbs and barbules, laterally meeting each other with their smooth surfaces creating a nearly air-tight seal, Wallace concedes to the Darwinists that this example shows the great importance of heredity, but it also presents one of the best examples of what Wallace called "directed power." In short, the bird's feather *is designed*. Wallace was particularly enthused about the design of the feather because of its microscopic intricacies and its macroscopic beauty—showing before our very eyes marvelous design at both levels.

Wallace believed insect metamorphosis to be another example. He paid special attention to Lepidoptera, whose change from a caterpillar into a mature butterfly he considered truly astonishing. The internal organs, sufficient for its life and growth as a caterpillar, dissolve then transform into "a perfectly different, and a much more highly organized creature." Yet he notes that from the humble beginnings of its larval form, the mature butterfly presents a display of color, pattern, and metallic beauty rivaling that of birds. Even more astonishing, he observes that unlike the bird's feathers which are essential to its survival, the coloration and patterns of the butterfly are "not functionally essential to the insect's existence."[187] Wallace admits that certain patterns and colors can have a protective purpose but he views the process as "unnecessarily elaborate." Why the whole process in the first place when the organic structure of the caterpillar seems to answer its needs? For Wallace the butterfly's metamorphosis was inexplicable by the mere principle of utility.

In either the case of the bird's feather or the butterfly, Wallace thought some other explanation than mere mechanistic processes were required. Building his case from multiple examples—the fine-tuning of the universe, the complexity of hemoglobin, and as already reviewed, the feather and metamorphosis of insects—he makes his bold declaration in chapter fifteen: "I now uphold the doctrine that not man alone, but the whole World of Life, in almost all its varied manifestations, leads us to the same conclusion—that to afford any rational explanation of its phenomena, we require to postulate the continuous action and guidance of higher intelligences; and further, that these have probably been working towards a single end, the development of intellectual, moral, and spiritual beings...."[188]

In chapter sixteen Wallace shows how general adaptations often go beyond the principle of utility and even the plant kingdom is uniquely suited to man's use. These facts too suggest a teleological world for Wallace. Wallace takes the opportunity to defend against such a view as "unscientific" by pointing out that he deduced design and purpose in nature from some of Darwin's own *descriptive* statements and, while naturalistic principles may *represent* certain phenomena, they by no means *explain* them. Wallace calls the assertions of Haeckel concerning an alleged unconscious "soul-atom" and similar speculations "vague and petty suppositions" that "do not meet the necessities of the problem."[189]

Wallace was not quite finished. He knew that in order for his "World of Life" to be a comprehensive theory he would need to deal with two related matters, one of which Darwin failed to address at all (the origin of life) and the second (the problem of pain and suffering) he addressed wholly inadequately. These two questions Wallace deemed of sufficient importance to devote two separate chapters.

In a broad sense the origin of life forms something of a leitmotif for the entire book. Wallace was not the first to notice its conspicuous absence in Darwin's work. Heinrich Bronn, who gave German readers their first introduction to *On the Origin of Species* with his critical translation in 1860, chided Darwin for not addressing this question. Why, asked Bronn, discuss the origin of species without addressing the more fundamental question of the origin of life? Until this was done Bronn considered Darwin's ideas hopelessly ambiguous.[190]

Darwin dismissed Bronn's criticism by referencing Leibnitz's objection to Newton's law of gravity on the grounds that Newton could not show what gravity itself precisely is.[191] But this was disingenuous. Surely Newton could clearly *demonstrate* its actions in measurable and repeatable ways. Newton, for example, could show that a body should fall to the earth at 3600 x 0.0044, or about 16 feet per second; this is a calculable and observable phenomenon. Whether it is the fall of an object to the earth or the sweep of the moon in its orbit, both are measurable, predictable, and due to the same force—gravity.

In contrast Darwin had nothing comparable. All Darwin could do was point to fossils and draw sweeping inferences that even *he* admitted were woefully scant and incomplete and insist that they *all* were the result of *random* variation. Darwin's "proofs" were inferential, not even evidentiary, much less predictable or repeatable. But if, as Darwin insisted, all life was linked through random evolutionary processes, Bronn's question of tracing this back to life's origin loomed large. All Darwin could do was dream of finding "some warm little pond" that might have produced primordial life from purely chemical processes.[192]

The problem of the origin of life was, for Wallace, essentially a problem of the cell. Could inorganic matter move to the structure and complexity of primitive first-life forms? This he addressed in chapter seventeen, "The Mystery of the Cell." Wallace dismissed the notion that life could have emanated from the mere accretions of protoplasm. Moreover, he attacked Huxley's notion of life as its own organizing power as a useless tautology, and referred to Haeckel's speculation of an unconscious "cell-soul" as a mere "verbal suggestion."[193] Force or matter, Wallace insisted, is inadequate to the task of explaining life's origin.

Reproducing Weismann's diagram of cell division, Wallace illustrated the profound complexity of the cell. Intricate sequencing of the membranous change, chromatin arrangement, division of the chromatin elements into equal parts, the appearance at opposite poles of centrosomes surrounded by a "sphere of attraction," the arising of delicate fibers or threads that pull the chromosomes with the disappearance of the nuclear membrane, the chromatin arrangement then becomes fixed, and finally the splitting longitudinally from "forces acting on the rods themselves" with the division completed by the two halves slowly drawn apart to the opposite poles approaching the center of attraction (the centrosome); all this takes place not by force, not by self-organizing power, or by a cell-soul, but *by design*. It is, in effect, a *directed cause*. The attempts of Haeckel and others to minimize these "marvelous powers" as the mere operations of chemistry Wallace called "wholly unavailing" and "mere verbal assertions that prove nothing" because they leave "All questions of antecedent purpose, of design in the course of development, or of any organising, directive, or creative mind as the fundamental *cause* of life and organization… altogether ignored…."[194]

To the notion that perhaps life and matter are eternal, Wallace replied that even assuming this theoretical construct riddled with massive problems, eternal life would simply suggest eternal life forces and energies directing and designing an eternity of progress. In essence, if life as exhibited in the progression of *Homo sapiens* is an example of what can happen over the course of a comparatively short period of time imagine time expanded to infinity! For Wallace the progressive development of biological life was neither a random nor a chance occurrence and since this progress obviously took place over time, erasing time constraints with a presumed infinity only served to magnify the intent and design of this development.

But where is this progress? Hadn't Darwin removed the pretensions of the natural theologians by noting the pervasive pain and suffering in this world? Wallace answered in chapter nineteen, "Is Nature Cruel?" Wallace notes the materialists' charge that no supreme intelligence would ever have created a world so wracked with pain and misery. Here Wallace invokes the principle of utility again, noting that since "no organ, no sensation, no faculty arises *before* it is needed, or in a greater degree than it is needed... [then] we may be *sure* that all the earlier forms of life possessed the minimum of *sensation* required for the purposes of their short existence; that anything approaching to what we term 'pain' was unknown to them."[195] Thirty years later C. S. Lewis would echo this same point.[196]

As for people, Wallace reminds the reader that pain has a purpose in protecting against threats to life and wellbeing. Pain is *necessary*. Physicians, who dealt with pain daily, had long understood this.[197] We might at this juncture legitimately ask about a broader issue related to pain, namely, what about human misery? Wallace knew misery. He knew in ways Darwin never did the financial difficulty in providing for a wife and family. But, like Darwin, Wallace also knew the loss of a child. Darwin, as we have seen, lost little Charlie at a very stressful time of his life, but he wasn't the first. Annie's death in 1851 is alleged by some to have been an important factor in what has been characterized as his "quiet disengagement from religious belief and spirituality."[198] Perhaps, but if the death of a child evoked similar responses from parents, the churches would have long since emptied. Infant and child mortality was all too commonplace through most of human history; the Victorian period was no exception.

The death of Wallace's beloved "Bertie" on April 24, 1874, (not quite seven years of age) of scarlet fever was surely a blow too. By then Wallace was a thoroughgoing spiritualist and amidst the death of his son he wrote "A Defense of Modern Spiritualism."[199] He also attempted a spirit communication with his son, but while assuaging the loss there is little doubt that the death of Bertie was a spiritually (*not* simply spiritualistically) affirming event. In contrast, Darwin's loss of 10-year-old Annie on April 23, 1851, while certainly felt no less than that of Wallace, confirmed his unbelief. Darwin never handled pain and adversity well, and as mentioned earlier, was sick and beset by psychological disorders most of his life. Darwin simply found suffering a by product of the vicissitudes of materialistic chance; for Wallace, used to privation and inured to struggle, pain was a necessary and sometimes instructive thread woven into a complex fabric of life.

In the final chapter of his *World of Life*, "Infinite Variety the Law of the Universe," Wallace, in another issue which profoundly separated him from Darwin, addressed an epistemological question—how *knowable* is an intelligent First Cause? Darwin despaired of an answer or more accurately answered negatively. Wallace, however, thought that he could provide at least a partially affirmative answer. As such, his *World of Life* became *his* "one long argument" in reply to Darwin's materialism. His argument is not specifically Christian, but even Wallace could not refrain from biblical allusions in his final summary. In the end, for Wallace, the sanctity of human life, so vacant in the writings of Charles Darwin, was an evident truth. "Man himself," he concluded, "[was] at his best, already 'a little lower than the angels,' and, like them, destined to a permanent progressive existence in a World of Spirit."[200]

10.

FINAL DAYS

LIFE AT BROADSTONE, 1902–1913

WALLACE WAS NEARLY 88 WHEN *THE WORLD OF LIFE* WAS PUBlished. He had made his grand evolutionary statement. Meanwhile he had turned his attentions to politics. A committed socialist who never forgot the poor farmers of his youth thrown off their land by a land enclosure act that benefitted the few over the many, Wallace supported land nationalization but rejected the Marxist call for government interference with labor and industry.

In some senses Wallace's "socialism" bears the marks of libertarianism. Ever sensitive to the meddling of the haves over the lives of the have-nots, Wallace rejected the "progressive" ideas of the eugenic movement that called for a program of social improvement promoting the "fit" over those deemed "unfit." First proposed by Darwin's cousin, Francis Galton, the eugenic applications of "social Darwinism" quickly transformed into a movement. By the early twentieth century, eugenics was reaching its height. In America, states were already passing eugenic laws that would sterilize without consent more than 60,000 citizens deemed mentally or morally "unfit."

Wallace's reaction to eugenics was closely defined by his scientific and political views. Wallace believed in free and open marriage unfettered by artificial constraints and obstructions. Social reform *not* human engineering would answer the question of progress. "Clear up [and] change the environments so that all may have an adequate opportunity of living a useful and happy life," he declared, "and give women a free choice in marriage" only then will you be in a position to guess which the "better stocks" are. As it was, eugenics served only as "a mere excuse for establishing a medical tyranny. And we have enough of this kind of tyranny already," he complained, "… the world does not want the eugenist to set it straight…. Eugenics is simply the meddlesome interference of an arrogant scientific priestcraft."[201]

Interestingly at Wallace's address celebrating the fiftieth anniversary of the joint reading of the natural selection papers by the Linnean Society (The Darwin-Wallace Celebration held at the Society on July 1, 1908), he told an illustrious audience that included eugenicist Galton, "I have long since come to see that no one deserves either praise or blame for the *ideas* that come to him, but only for the *actions* resulting therefrom…. I therefore accept the crowning honour you have conferred on me to-day, not for the happy chance through which I became an independent originator of the doctrine of 'survival of the fittest,' but, as too liberal recognition by you of the moderate amount of time and work I have given to explain and elucidate the theory, to point out some novel applications of it, and (I hope I may add) for my attempts to extend those applications, even in directions which somewhat diverged from those accepted by my honoured friend and teacher—Charles Darwin."[202] It is hard to miss the stab at Galton's social Darwinism—"the *actions* resulting therefrom"—here, and it is one of the few times Wallace took pains to distinguish himself from Darwinian evolution by mentioning his "novel applications."

Similarly, Wallace thought the public health campaign to mandate vaccination against smallpox was simply another example of governmental intrusion into the lives of its citizens, an intrusion the purpose of which remained unproven. In retrospect Wallace's stand seems unreasonable and even dangerous, but as public health biologist Thomas P. Weber has convincingly demonstrated, the statistics used by Wallace and the vaccinationists were based on actuarial figures ill-suited to resolving the question of the procedure's efficacy.[203]

Weber points out that it is unfair and unhistorical to simply paint Wallace as a religiously motivated crank. "In the case of vaccination," writes Weber, "Wallace argued that liberty and science need to be taken into account, but that liberty is far more important than science. Wallace only appears to have been such a heretical figure if a large portion of the social, political, and intellectual reality of Victorian and Edwardian England is blotted out of the picture.... The Victorian vaccination legislation was part of an unfair, thoroughly class-based, coercive, and disciplinary health care and justice system: poor, working-class persons were subjected to the full force of the law while better off persons were provided with safer vaccines and could easily avoid punishment if they did not comply."[204] Indeed Wallace's conclusions based upon the actuarial figures were not illogical, they were just based on poor data as were those of his opponents, and yet the debate itself was instrumental in the development of sound inferential statistics that would finally settle the question.

Even as he approached ninety, Wallace remained active mentally and physically. In 1902 he designed and moved into Old Orchard at Broadstone, Dorset. His last two books were *The Revolt of Democracy* and *Social Environment and Moral Progress*, the latter containing an extended blast against the eugenicists. Comfortably settled into a routine at Old Orchard he and Annie enjoyed their senior years. As late as February of 1913 Annie could write, "Dr. Wallace is very well and busy, writing as hard as ever; he has just passed 90 and feels like 50."[205] Despite comparatively good health, the sand was about to run out of Alfred's hourglass. At 9:25 on the morning of November 7, 1913, nature's adventurer departed this world. The symbolism of Wallace's last works in comparison to Darwin's is hard to ignore. Wallace spent his final years seeking to synthesize his philosophical and political views, while Darwin produced a book on worms. In the end, Wallace's final efforts were prolegomena for his eternity as indeed were Darwin's.

"OLD ORCHARD," BROADSTONE
(Built in 1902)

11.

WALLACE'S LOST LEGACY

ERASING A LIFE

IT IS HARD TO REDUCE SO ACTIVE AND VARIED A LIFE AS ALFRED Russel Wallace's to a single legacy. Surely his scientific contributions in evolutionary theory, natural selection, biogeography, anthropology, or zoology have long been chronicled and discussed.[206] Less known and often mischaracterized are his efforts at establishing a natural theology explicitly incorporating evolutionary theory.

Interestingly, Wallace was very much a part of that discourse among his contemporaries. John Magens Mello, vicar of Mapperly and fellow of the Geologist Society, was captivated by *The World of Life*. He saw no problem with having the natural world guided by intermediary beings. "To whatever extent any may be disposed to accept or reject these views [of Wallace's] upon Creation, we must all of us admit, if we do not set aside the teaching of Holy Scriptures, that there are in the Universe Spiritual Intelligences besides Man; Beings over and over again referred to in the Bible; and we are here taught that by God's appointment they have special duties and work to perform in connection with this World and with us Men. Our Lord Himself speaks to us in no uncertain terms of the Ministry of Angels, and of the interest they take in Human life."[207]

Wallace's ideas even influenced the birth of modern Christian fundamentalism, itself a movement often misunderstood today. The fundamentalist movement began in 1909 under the patronage of two American laymen Milton and Lyman Steward. Together they commissioned a series of essays designed to form a modern apologetic of Christian fundamentals. Those articles were subsequently compiled and published in 1917 as *The Fundamentals: A Testimony to the Truth*.

One of those contributors, the Reverend and Professor James Orr of the United Free College, Glasgow, Scotland, wrote "Science and Christian Faith," which included a spirited defense of evolution and Christian theism.[208] "In truth," wrote Orr, "no conception of evolution can be formed, compatible with all facts of science, which does not take account, at least at certain great critical points, of the entrance of *new factors* into the process we call creation."[209] Those points, Orr went on to say, were "the transition from inorganic to organic existence," the beginning of the "development of *consciousness*," and finally "the transition to *rationality, personality*, and *moral life* of man." These were each creative acts attributable to a higher force acting *upon* nature. This was pure Wallace.

Unfortunately, when the fundamentalist movement took more organized form at a 1919 conference in Philadelphia, Orr's perspective was cast aside. No longer was it sufficient to hold to the infallibility of Scripture, the deity of Christ, Christ's atonement, and the return of Christ at a second coming, now a young earth and complete rejection of *any* form of evolutionary theory were added. It may well be said that Wallace's natural theology was expunged from the Christian fundamentalist movement at that meeting.

It is unfortunate that something for which Wallace became so strongly devoted has been lost or worse, caricatured. The comparatively unexplored aspect of Wallace's role within the larger context of natural theology represents a complex omission. In part, as just mentioned, it was the product of ill-considered religious balkanized politics. But it is also related to Wallace's secondary status in the history of biology and science. Darwin has so overshadowed the historical discourse that Wallace is often recast in the image of the dominant co-discoverer of natural selection.

In spite of Wallace's extensive work in developing an overtly teleological and theistic evolutionary theory, few have examined it in detail. When discussed at all, historians have generally regarded Wallace's metaphysical views in one of three ways: 1) Wallace was an *ultra-naturalist*—there *is* no theism in Wallace's worldview; 2) Wallace was a *spiritualist* and his larger worldview is interpreted solely through this spiritualist lens, leaving any teleological or theistic implications largely unaddressed; or 3) Wallace developed a cosmology and biology that was *rooted in teleology and theism.* Because each historiographical school has influenced interpretations of Wallace's life and work, each needs to be examined with some care. These will be taken up in order of prominence.

The most prevalent view is the one that sees Wallace as an ultra-naturalist/materialist. The irony is that this perspective is shared by very different subgroups. Creationists otherwise sympathetic to Wallace's world of purposeful guidance fault him for supposedly not being a theist.[210]

At the same time, those who give *a priori* privilege to methodological naturalism in scientific discourse embrace Wallace as one of their own. The latter group's commitment to methodological naturalism, as we shall see, profoundly influences their interpretation of Wallace's biology and cosmology. Champions of this idea include Michael Shermer, Steven J. Dick, and Charles H. Smith. Shermer sees Wallace as adhering to philosophical scientism. Standing the Wallace-as-spiritualist thesis on its head, Shermer argues that "the causal vector was in the other direction. Wallace's scientistic worldview forced him to shoehorn his encounters, experiences, and experiments in spiritualism into his larger scientism."[211]

If scientism is given its ordinary definition, namely, the notion that the investigational methods of the natural sciences should be applied in *all areas* of inquiry, then Shermer clearly belongs to the ultra-naturalist historiographical stream of Wallace scholarship. For Shermer, Wallace's all-encompassing law-based scientism was a product of his "heretic personality."[212]

Steven J. Dick's analysis of *Man's Place in the Universe* argues for Wallace's "anthropocentric and teleological world view" (understood within the context of an anthropic principle) by linking his cosmological ideas to those of "self-proclaimed atheist" Fred Hoyle.[213] While versions of the anthropic principle can accommodate theism, Dick suggests that

Wallace took a decidedly non-theistic view by making "habitability" the self-directed "goal" of the universe. Thus, under this suggestion, Wallace's views broadly comply with James Lovelock and Lynn Margulis's Gaia hypothesis. Smith largely agrees, though with some revisions. For Charles H. Smith, Wallace anticipated cybernetics in developing a thoroughly law-based, nonteleological "evolution as a spatial interaction process."[214] Here Smith makes much of Wallace's likening natural selection to a "governor on a steam engine."[215] Wallace's worldview was driven by final cause, he claims, but he rejects its teleological implications, suggesting, "We need not adopt the *more extreme* [emphasis added] of these [teleological views] to suggest how a system as described here could find its way to higher order...."[216]

There are problems with each of these arguments. Shermer's application of "heretic personality" is inconsistent and vague. As John van Wyhe pointed out, Shermer's effort amounts to little more than a collection of "numerous subjective impressions" and his "appeal to a heretic personality" becomes simply a "redescription of details of Wallace's interests. No new insights are offered by the heretic personality thesis."[217]

As for Dick, Hoyle was *not* an atheist later in life. While not a Christian, Hoyle rejected purely chemical explanations for the origin of life, was a staunch critic of Darwinian evolution, and believed that the "information-rich" universe was controlled by an "overriding intelligence."[218] He rejected the "crude denial of religion... prevalent among so-called rationalists of the late nineteenth century,"[219] and insisted, "The atheistic view that the Universe just happens to be here without purpose and yet with exquisite logical structure appears to me to be obtuse...."[220] In fact, Dick's association of Wallace with a self-directed anthropic principle or Gaia hypothesis is simply arbitrary and not supported by his own example.

Smith's argument is no better. He fails to explain why a theistic teleology should be considered a "more extreme" view, while his cybernetic connection with Wallace is anfractuous, presentist and speculative. Smith's essential problem with associating Wallace's views with cybernetics is that Norbert Wiener's theory, a term coined in the summer of 1947, is ill-suited to explaining *intelligence* or mind, the very thing necessary to cohere to Wallace's frequent references to "Overruling Intelligence," "Mind," or "Infinite Being."[221] While Smith appreciates the need to understand information as "part and parcel of organized adaptive structure" and as potentially supportive of "new-kinds of information-sharing networks at the ecological/environmental level,"[222] this kind of *quantitative* Shannon information is inadequate to explain the complex, specified information necessary to account for Wallace's requisite *intelligence*.[223] Finally, Smith also probably makes too much of the "governor on a steam engine" phrase. That Wallace was glimpsing cybernetics is doubtful.

More likely is Muriel Blaisdell's point that grand analogy was commonplace among natural theologians of the period who made frequent reference "between products of Divine and human manufacture."[224] Wallace does this very thing in *The World of Life* when he writes, "the Mind which first caused these elements to exist, and then built them up into such marvellous living, moving, self-supporting, and self-reproducing structures, must be many millions times greater than those which conceived and executed the modern steam-engine."[225]

The *real* problem with all these conjectures is that they do not comport to Wallace's own views. One gets the sense that there are more of the authors than their subject in these "ultra-naturalist" claims. Each of them try to cast Wallace into a mold to which he never conformed; each appear to hold *a priori* commitments to methodological naturalism, commitments that preclude cosmologists from adopting teleological and theistic fine-tuning hypotheses for the origin of the universe or biologists from invoking information theory premised upon an understanding of life as intelligently engineered.

When those hidebound to methodological naturalism *do* apply fine-tuning or information-based arguments, they are normally cast as non-teleological anthropic principles or some self-sustaining Gaia hypothesis. But, as William Lane Craig has pointed out, "What is striking about methodological naturalism is that it is a philosophical, not a scientific, viewpoint. It is not an issue to which scientific evidence is relevant; it is about the philosophy of science. As such, it is notoriously difficult to justify."[226]

In keeping with their methodological prejudices Shermer, Dick, and Smith emphasize Wallace's *law-based* universe; all of them are fond of referencing Wallace's rejection of first cause in nature.[227] However, as discussed earlier, Wallace simply rejected the idea of a first cause in nature *not* a First Cause. Smith believes that "we have yet to *prove*"[228] Wallace's theism, and yet its best proof comes from Wallace himself. In a letter to his close friend and biographer Reverend James Marchant written just before his death, Wallace spelled out succinctly and in no uncertain terms his natural theology:

> The completely materialistic mind of my youth and early manhood has been slowly moulded into the socialistic, spiritualistic, and *theistic mind* [emphasis added] I now exhibit—a mind which is, as my scientific friends think, so weak and credulous in its declining years, as to believe that fruit and flowers, domestic animals, glorious birds and insects, wool, cotton, sugar and rubber, metals and gems, were all foreseen and foreordained for the education and enjoyment of man. The whole cumulative argument of my "World of Life" is that *in its every detail* it calls for the agency of a mind... enormously above and beyond any human mind... whether thus Unknown Reality is a single Being and acts everywhere in the universe as direct creator, organizer, and director or every minutest motion... or through "infinite grades of beings", as I suggest, comes to much the same thing. Mine seems a more clear and intelligible supposition... and it is the teaching of the Bible, of Swedenborg, and of Milton.[229]

The leap made by the ultra-naturalists from Wallace's law-based nature to a rejection of theism comes from a misunderstanding of nature's laws and of science. Christian apologist C. S. Lewis pointed out, "*in the whole history of the universe the Laws of Nature have never produced a single event....* All events obey them, just as all operations with money obey the laws of arithmetic.... But arithmetic by itself won't put one farthing into your pocket."[230] These laws, he notes, are "an empty frame" into which events and actions need to be placed. Put another way, "Natural laws do not account for the origin of all events any more than the laws of physics alone explain the origin of an automobile. Natural laws account for the *operation* of these things."[231]

Wallace understood what Shermer, Dick, and Smith seem not to, namely, that the *action* of laws need an explanation, thus he posited *subordinate* entities (often referring to them as "angels") to act upon these laws. Related to this are miracles and the laws of nature. Shermer and Smith object to miracles on largely Humean grounds. But there is no reason to assume that Godly intervention must necessarily "suspend" or "break" the laws of nature.[232]

A more fundamental question, however, is, does a "law-based" nature preclude miracles and an adherence to scientific method? More specifically, does Wallace's acceptance of what some would call the "miraculous" somehow taint his science? The question might be turned: Why must scientists preemptively reject miracles? "Scientists, as scientists," Norman Geisler observes, "need not be so narrow as to believe that nothing can ever count as a miracle. All a scientist needs to hold is the premise that every event has a cause and that the observable universe operates in an orderly way."[233] Thus, a law-based nature tells us little about divine agency; in fact, it simply begs the theistic question.

In the end, the ultra-naturalists commit the same fallacy as Romanes. They take an aspect of Wallace's theory and twist it into a counterfeit mutation of its original form. Shermer turns Wallace's devotion to scientific inquiry into scientism; Dick strips Wallace's cosmology of its teleology and theism transforming it into a kind of proto-Gaia universe; and Smith confuses Wallace's guided and directed intelligent evolution with a murky cyber-spatial world of naturalistic final cause. All are disconcertingly presentist; all would have surprised and puzzled Wallace.

The second historiographical current in relation to Wallace is the one that seeks to explain him by and through his spiritualism. Under the spiritualist category Malcolm Jay Kottler argues for two Wallaces a pre-spiritualist Wallace and a post-spiritualist Wallace.[234] For Kottler, all of Wallace's discussions of "Overruling Intelligence," "Infinite being," or "guiding spirits" emanated from his conversion to spiritualism.

The problem with this view is that Wallace's teleological and theistic leanings are discernable well *before* his first séance. First agreeing with most skeptics that the Gospel miracles were myths, he later recanted, admitting that such doubts were "based upon pure assumptions which were not in accordance with admitted historical facts."[235] Nevertheless, even during this early period there is reason to believe he harbored some religious inclinations. As a young man of twenty writing late in 1843 he asked, "can any reflecting mind have a doubt that, by improving to the utmost the nobler faculties of our nature in this world, we shall be the better fitted to enter upon and enjoy whatever new state of being that future may have in store for us?"[236] This linkage of humanity's "nobler faculties" with a progressive impulse toward a "new state of being" would form the heart of his evolutionary teleology later in life.

Similar views occasionally came out in Wallace's early scientific writings as in his, "On the Habits of the Oran-utan of Borneo." Noting that this remarkably human-like beast had many physical attributes not specifically useful for survival, he suggested other examples in nature, concluding with no little irritation that "Naturalists are too apt to *imagine*, when they cannot *discover*, a use for everything in nature: they are not even content to let 'beauty' be a sufficient use, but hunt after some purpose to which even *that* can be applied by the animal itself, as if one of the noblest and most refining parts of man's nature, the love of beauty for its own sake, would not be perceptible also in the works of a Supreme Creator."[237] Even early on Wallace was preconditioned to construe natural selection as a *limited* explanatory mechanism with nature perhaps having a divine aesthetic intent; this two years *before* the famous Ternate letter.

Spiritualism was not Wallace's religion; Wallace's religion incorporated spiritualism. In fact, as we have seen, his break with Darwin came from his reading of the science *not* from his metaphysic. Ross A. Slotten similarly makes much of the influence of Swedenborgian mysticism,[238] and while Wallace was surely influenced by Swedenborg's cosmology of corresponding spirit realms, it can easily be overdrawn. Wallace himself never publicly proclaimed his allegiance to Swedenborg. While spiritualism may be suggestive of a non-materialistic and decidedly *un*naturalistic metaphysic, it was in no way dictated *by* it.

There is yet another reason to reject spiritualism as a defining force in Wallace's metaphysic. Wallace pointed out that no orthodox religious belief, or *any* particular theistic belief (Catholic, Protestant, Islamic, or Hindu), was ever confirmed in a legitimate séance. Wallace explained that every time a question about God or Christ was asked the spirits themselves never relayed more than an opinion or, more often, the admission that they had no more knowledge of the divine than they had in life.[239] Thus, these departed souls were not related to Wallace's elaborate discussions of "Overruling Intelligence," of "Mind," of "Infinite Being," of "angels," or of "intelligent beings." Wallace *never* used these terms in connection with spirit communications in or out of a séance, and they are not in any meaningful sense connected to the departed souls integral to his belief in spiritualism. They really *are* different entities.

Wallace was, if nothing else, independent, a man keenly attuned to the *Zeitgeist* of the age who absorbed ideas and then synthesized them into his own unique worldview. While the impact of spiritualism on Wallace (and many others in Victorian society) was profound, care should be taken in defining the precise nature of that impact.

Spiritualism lodged itself into nineteenth-century English class structure differently. Wallace was deeply influenced by what Logie Barrow has called "plebian spiritualism," a working- and lower-middle-class arm of the movement that initially tended toward more radical political activism, unorthodox religious beliefs, and even avowed hostility toward Christianity.[240] But plebian spiritualism changed. Many of its proponents eventually abandoned their former secularism, came to regard materialism as "boring and 'negative,'" and pursued various spiritual, Christian and quasi-Christian outlets as more satisfying.[241] Wallace surely fits this general pattern of plebian spiritualists. Nevertheless, centralizing spiritualism into a defining theme in Wallace's teleology, as Kottler does, amounts to the tail wagging the dog.

Unfettered by such *modus operandi*, fresh perspectives can be applied, which brings us to the third view of Wallace's metaphysics. Martin Fichman of York University is rare but perceptive in his extensive discussion of Wallace as a theist when he notes, "Few have adequately examined Wallace's broader religious worldview, his evolutionary theism."[242] Referring to *The World of Life*, Fichman writes, these were "not the eccentric musings of a declining mind but powerful syntheses of late-nineteenth/early twentieth-century intellectual currents."[243]

Given the problematic nature of the spiritualist and non-theistic arguments, the conclusion seems clear enough: Wallace was not only a theist but devoted considerable attention to refurbishing natural theology along what he felt were more robust scientific lines. Martin Fichman is right: "the *World of Life* was written to demonstrate that the most recent scientific researches rendered natural theology (in sharp contrast to revealed theology) both reinvigorated and essential for the twentieth century."[244]

It remains to examine Wallace's natural theology itself. The brief but unequivocal declaration to Marchant quoted earlier (see page 104) in this chapter gives a fascinating and insightful clue as to the sources of its construction. Most interesting is Wallace's appeal to the Bible. While Wallace always rejected certain Christian doctrines (particularly sin, judgment, atonement, and damnation), he never strayed far from biblical references in his metaphysical utterances. For example, alluding to Psalm 8:5 he refers to, "Man himself... 'a little lower than the angels,' and, like them, destined to a permanent progressive existence in a World of Life."[245]

In fact, Wallace had little problem with biblical terminology in his theology: "I believe all this [natural world] to be [under] the guidance of beings superior to us in power and intelligence. Call them spirits, angels, gods, or what you will; the name is of no importance. I find this control in the lowest cell. The wonderful activity of cell life convinces me," he declared in an interview, "that it is guided by intelligence and consciousness. I cannot comprehend how any just and unprejudiced mind, fully aware of this amazing activity, can persuade itself to believe that the whole thing is a blind and unintelligent accident."[246] It also cannot go unnoticed that Wallace's support for theism is derived from the natural complexity of the cell, an argument currently common among intelligent design advocates.[247]

Wallace's reference to Swedenborg, with its psychic accounts of spirit-being realms and progressions, was obviously of considerable influence, but his final mention of Milton is more intriguing. Milton suggests a Christian connection drawn from premodern theology: the nine heavenly orders described in the celestial hierarchy. Attributed to Dionysius the Areopagate (*circa* 5th century), the idea of an ordered ranking of angels "whose obedience and ministry God employs to execute all the purposes which he had decreed" was taken up by Thomas Aquinas and numerous divines for over a thousand years.[248]

Wallace's mention of Milton is interesting in this regard since the seventeenth-century poet was one of the last to extensively acknowledge that the angels are "distinguisht and quaterniond into their celestiall Princedomes and Satrapies."[249] It is hard not to see the celestial hierarchy at work in Wallace's theology. Again, Wallace was no Christian. Rejecting man as a fallen and sinful creature, Wallace had no need of redemption or a Redeemer. Yet Wallace never strayed far from the Bible and its *scala naturæ*. Wallace was an idealist not a mystic; he rejected the panentheism of Henri Bergson and scoffed at theosophy. While Martin Fichman thinks that Wallace was a "precursor of twentieth-century process theology,"[250] his correspondence suggests otherwise. Bergson, to whom Alfred North Whitehead acknowledged a tremendous debt, left Wallace unimpressed.

In a letter to Oxford University's Hope Professor of Zoology, Edward Bagnall Poulton, dated May 28, 1912, the aged Wallace admitted to not having read Bergson but added that from what he understood, the French philosopher's "vague ideas" such as "an internal development force" seemed to him "of no real value as an explanation of Nature." For Wallace the Overruling Intelligence or Mind worked "by and through the primal forces of nature" not *in* them and certainly not, Wallace added, by some vague "law of sympathy." Wallace confessed to Poulton that he didn't think he could read such a book.[251]

Drawing from Christian and spiritualist sources, Wallace's cosmology and biology were important if now-forgotten counters to the rising tide of Victorian materialism, supporting in perhaps an unexpected way more recent scholarship suggesting that the angelic hierarchy lived on past the Reformation.[252] Thus, Wallace presents a Janus-faced view. Looking to science for a new and powerful foundation upon which to build a natural theology for the future, he filled in those details by taking, in several senses, a premodern approach.

This hardly makes Wallace *pro mortuus* in the history of ideas. It may, in fact, be very much alive in the current debate over the nature of science and the biological paradigm. William A. Dembski, a leading intelligent design theorist, has noted that while much of premodern thought was worth discarding, only premodernity entails "a worldview rich enough to accommodate divine agency."[253] So in a key sense Wallace becomes a pivotal figure between past and present. David Kohn has said, "Darwin, the last of the natural theologians, is the man who turned out the lights."[254] If indeed Darwin turned off the lights to natural theology, Wallace surely tried to turn them back on.

In a very real sense Wallace was a traditionalist who sought to bring science back to its more expansive moorings. Like most of the great scientists before him, Wallace believed in the uniformity of natural causes. What Darwin gave the world, unlike anyone before him, was a uniformity of natural causes *in a closed system*, a system strictly bounded between what Francis Schaeffer called the "upper story" of grace and spirit and the "lower story" of nature and law. Indeed as Schaeffer points out the closing off of these two spheres into a NOMA (non-overlapping magisteria)[255] "makes all the difference in the world. It makes the difference between natural science and a science rooted in naturalistic philosophy."[256] Wallace's goal was to reunite what he viewed as a fractured worldview burdened by naturalistic philosophy into a more holistic worldview enriched by natural theology.

If much of Wallace's legacy was lost by a coincidental alliance of religious zealotry and secular theorists, it was also buried by whiggish evolutionary biologists and historians. Whiggish history is a term coined by Herbert Butterfield. It is "the tendency in many historians [and others]... to praise revolutions provided they have been successful, to emphasise certain principles of progress in the past and to produce a story which is the ratification if not the glorification of the present. This whig version of the course of history is associated with certain methods of historical organisation and inference—certain fallacies to which all history is liable...."[257]

If all history is prone to this presentist fallacy it has been made manifest in regard to Alfred Russel Wallace. A few examples will suffice. John van Wyhe in his essay review of Slotten's biography of Wallace, for example, insists that Darwin deserves to be remembered over Wallace precisely *because* the latter disagreed that domestic breeding examples shed any light on the process of natural selection, and, moreover, that his suggestion that man was unique and separate from animals was rendered "without any evidence." He then proceeds to Wallace-bash by further claiming that unlike his counterpart he "did not spend more than 40 years scouring the literature for relevant information."[258]

Of course van Wyhe does *not* point out that Wallace had more than twice the field experience of Darwin nor does he point out any of the things discussed earlier in this biography that brought Wallace to his conclusions regarding the nature of man. This is classic whig history. Interestingly, in response to charges of whiggishness in his own portrayal of Darwinian evolution, the late Ernst Mayr offered special pleadings and then attempted to debunk Butterfield's concept.[259] "When is historiography whiggish?," Mayr asked. Apparently not when Mayr engages in it or when he *does* to dismiss the term as "ill advised" and not appropriate to the history of science. Whiggishness it seems is a charge to which Darwinians are both prone and sensitive.

All of this suggests that Wallace needs to be placed within a more coherent context, one that has value in framing the ongoing discourse in biology and cosmology. While the Darwinian paradigm remains contested, understanding the co-discoverer of natural selection as the founder of an evolutionary theory deeply rooted in natural theology gives the debate a more accurate historical reference point. Instead of the reductionist "science versus creationist" arguments, much would be improved by a more nuanced approach. That approach would factor in the historical dynamics bearing upon an all-too-polarized dialogue between those equating a strict methodological naturalism (a philosophical presupposition) with science (an inquiry for truth in the natural world) and those suggesting a counterview that was, in many important ways, anticipated by Alfred Russel Wallace. Regardless of one's place or position in this current dialogue, it has too often taken place with little or no appreciation for the historical landscape in which it was formed.

One last contemporary interpretation of Wallace deserves mention. It comes from evolutionary psychologist Steven Pinker. Pinker fully admits to Wallace's belief in intelligent design and even goes so far as to call him a creationist. As we have seen, Pinker is not far wrong in his assessment of Wallace. However, Pinker claims that Wallace deserves to be dismissed because *he* (Pinker) has solved the problem of the human mind and evolution by purely naturalistic means. He claims to have done so with two concepts. First, what he calls "the cognitive niche." This is "a mode of survival characterized by manipulating the environment through causal reasoning and social cooperation." Second, a theory of co-option whereby "the cognitive niche can be co-opted to abstract domains by processes of metaphorical abstraction and productive co-operation, both vividly manifested in human language."[260] The problem with Pinker's "solution" is that it is entirely speculative and not based upon any large body of evidence.

Furthermore, Pinker claims that the human mind was "designed by natural selection."[261] This is problematic in the extreme. In what sense does natural selection *design* anything? Edward Feser has noted this problem among those attempting a Darwinian and materialistic explanation for the mind:

> [T]he operation of the mechanism the theory appeals to in order to explain intentionality itself presupposes intentionality. That this criticism seems to apply to the biological theory as much as to the causal theory [of the mind] is even more evident when one considers that ultimately, there may be no substantive difference between them. For… the trouble with appeals to biological function in this context is that all talk about biological function must, from a Darwinian point of view anyway, be regarded as nothing more than shorthand for talk about causation. To say that the heart was selected by evolution to serve the function of pumping blood is, strictly speaking, to say something false; for evolution doesn't literally serve any purpose or function at least not in the Darwinian view. Indeed, *the whole point* of Darwin's account of evolution by natural selection is to get rid of the need to appeal to literal purposes and function in terms that make reference only to purposeless, meaningless causal processes. The right thing to say about the heart is, in a Darwinian view, just this: it causes blood to flow, and that it was in turn caused by a series of successive genetic mutations that allowed the creatures exhibiting them to survive and reproduce in greater numbers than those which lacked them. And that's it. If talk about the "purpose" or "function" for which the heart was "selected" has any application at all, it is only as a way of noting how what in reality are the purposeless, functionless, and meaningless results of unthinking causal processes can seem to be purposive, functional, and meaningful.
>
> Talk about purposes and functions, if taken literally, seems to presuppose intentionality; in particular, it seems to presuppose the agency of intelligence who designs something for a particular purpose or to serve a particular function. But the aim of Darwinian evolutionary theory [and Pinker's "cognitive niche"] is to explain biological phenomena in a manner that involves no appeal to intelligent design.… This is of a piece with the general materialistic tendency to regard genuine scientific explanation as requiring the stripping away of anything that

smacks of the subjective, first-person, intentional point of view. It thus seems odd that materialist philosophers [including Pinker] should think it a hopeful strategy to appeal to biological function in order to explain intentionality.[262]

Some have called into question Darwinists' efforts to explain the operations of the mind on still other grounds. Johann J. Bolhuis and Clive D. L. Wynne in a recent *Nature* article have pointed out that two decades of animal studies of cognition that impute monkeys with human traits such as empathy and conflict resolution through "a sense of fairness" and reconciliation lacked appropriate controls and have been subject to "a flurry of anthropomorphic overintepretation." They suggest that "[s]uch findings have cast doubt on the straightforward application of Darwinism to cognition. Some have even called Darwin's idea of continuity of mind a mistake."[263] Pinker does not strengthen his case with an appeal to human language. Bolhuis and Wynne have pointed out that researchers have tried for years to teach primates to use language and all have failed. "One of the prerequisites for language," they explain, "is being able to imitate sounds that are created by someone else. Our primate cousins show no inclination to do this. Yet many parrots and songbirds are striking vocal mimics.... The appearance of similar abilities in distantly related species, but not necessarily closely related ones," they conclude, "illustrates that cognitive traits cannot be neatly arranged on an evolutionary scale of relatedness."[264]

Thus it seems Pinker's "solution" remains unconvincing. Pinker's "cognitive niche" seems to bear out something David Berlinski has noted: "The largest story told by evolutionary psychology is therefore anecdotal. Like other such stories, it subordinates itself to the principle that we are what we are because we were what we were. Who could argue otherwise? All too often, however, this principle is itself supported by the counter-principle that we were what we were because we are what we are, a circle not calculated to engender confidence."[265]

While Pinker is absolutely correct in associating Wallace with intelligent design, his efforts to dismiss the naturalist's original call for something other than natural selection to account for the human mind falls far short of its mark. "Materialism may be the majority position in contemporary philosophy of mind," states Feser, "but not because anyone has proved it true. Indeed... virtually all work done today by materialist philosophers of mind consists, at bottom, of trying to defend their favored brands of materialism against various objections, which are implicitly or explicitly anti-materialist in character.... Moreover, these objections are typically variations on the same criticisms of materialism that have been given for 2,500 years, with modern materialists no closer to answering them decisively than were their intellectual forebears."[266] In short, a materialistic/Darwinian solution to the problem of the emergence of the human mind by evolutionary means remains as intractable today as ever. Perhaps that is because the only viable solution was proposed by Wallace more than 140 years ago.

Epilogue

Wallace and Modern Intelligent Design

WHERE ARE WE TO PLACE WALLACE WITHIN THE CONTEMPORARY intelligent design movement (ID)? Despite naysayers like Charles H. Smith (see his FAQ #1 @ Wallace Page at http://people.wku.edu/charles.smith/index1.htm)who insist that Wallace couldn't have been an ID proponent, Wallace presents some important links with the current movement. While Wallace could not have access to modern astronomy and genetic principles, his work seemed to glimpse an ID future. His cosmology is echoed in Guillermo Gonzalez and Jay Richards's *The Privileged Planet*, and his *World of Life* chapter "The Mystery of the Cell," anticipated Stephen Meyer's *Signature in the Cell*.

If anything, Wallace's position went well beyond the scientific proposition of ID, which simply states that certain features of the natural world give evidence of intelligence and design, without prejudging whether the source of that design is inside or outside of nature. Even a superficial reading of *The World of Life* suggests Wallace insisted that something from *outside* of nature itself was necessary for the creation of life. His interview "New Thoughts on Evolution" with Harold Begbie in 1910 anticipating the release of his book is revealing. (See Appendix C for the transcript.) Asked how life began on earth, Wallace said this:

> Well, it is the very simple, plain, and old-fashioned one that there was at some stage in the history of the earth, after the cooling process, a definite act of creation. Something came from the outside. Power was exercised from without. In a word, life was given to the earth. All the errors of those who have distorted the thesis of evolution into something called, inappropriately enough, Darwinism have arisen from the supposition that life is a consequence of organisation. This is unthinkable. Life, as Huxley admitted, is the cause and not the consequence of organisation. Admit life, and the hypothesis of evolution is sufficient and unanswerable. Postulate organisation first, and make it the ori-

gin and cause of life, and you lose yourself in a maze of madness. An honest and unswerving scrutiny of nature forces upon the mind this certain truth, that at some period of the earth's history there was an act of creation, a giving to the earth of something which before it had not possessed; and from that gift, the gift of life, has come the infinite and wonderful population of living forms. Then, as you know, I hold that there was a subsequent act of creation, a giving to man, when he had emerged from his ape-like ancestry, of a spirit or soul. Nothing in evolution can account for the soul of man. The difference between man and the other animals is unbridgeable. Mathematics is alone sufficient to prove in man the possession of a faculty unexistent in other creatures. Then you have music and the artistic faculty. No, the soul was a separate creation.

But are these the only two instances of interference from outside?

Ah, we come to a great question. I deal with it in a book which Chapman and Hall are to publish this winter. In some ways this book will be my final contribution to the philosophic side of evolution. It concerns itself with the great question of Purpose. Is there guidance and control, or is everything the result of chance? Are we solitary in the cosmos, and without meaning to the rest of the universe; or are we one in 'a stair of creatures,' a hierarchy of beings? Now, you may approach this matter along the metaphysical path, or, as a man of exact science, by observation of the physical globe and reflection upon visible and tangible objects. My contribution is made as a man of science, as a naturalist, as a man who studies his surroundings to see where he is. And the conclusion I reach in my book is this: That everywhere, not here and there, but everywhere, and in the very smallest operations of nature to which human observation has penetrated, there is Purpose and a continual Guidance and Control.

These comments go well beyond the scope of ID as a scientific theory. So while Wallace may justifiably be considered a precursor to modern ID, he should not be considered its founder. However, it should not be ignored that this co-discoverer of natural selection rejected Darwinian materialism in favor of an argument from design for biological life. In this sense modern evolutionary theory may be seen as emanating from two traditions: one from Darwin who established the current evolution paradigm rooted in methodological naturalism and philosophical materialism; the second from Wallace rejecting methodological naturalism in favor of an open inquiry to discover the truths of nature and further rejecting the myopic vision of materialism. Once the current debate is reconfigured to eliminate old unhistorical and whiggish rhetorical faultlines—"science" versus "superstition," evolution versus "creationism," etc.—Alfred Russel Wallace will become, after more than a century, a rediscovered life.

APPENDICES

THE THREE APPENDICES PRESENTED HERE ARE SIGNIFICANT EX-
amples not only of Alfred Russel Wallace's theory of evolution, but
of Wallace's own intellectual evolution.

Appendix A, "The Ternate Letter," is his initial theory of natural
selection sent to Charles Darwin in the winter/spring of 1858.

Appendix B is an excerpt from his chapter "Darwinism Applied to
Man" in his book *Darwinism*, which was published in 1889 and is an
elaboration of his dramatic departure twenty years earlier from Darwin's materialistic theory of evolution. More specifically it represents
Wallace's direct challenge to Darwin's *Descent of Man* (1871).

Finally, Appendix C, "New Thoughts on Evolution," is from an interview in which Wallace elaborates on his views of the origin of life,
teleology in the natural world, and their philosophical implications. The
source reprinted here is from a small chapbook issued by the London
publishing house Chapman and Hall to promote *The World of Life*,
Wallace's grand synthesis of evolution, intelligent design, and ultimate
purpose released in 1910.

A.

THE TERNATE LETTER

by Alfred Russel Wallace

Introduction by Michael Flannery

This is the famous letter that Charles Darwin said he received unexpectedly on June 18, 1858. Shocking Darwin with a suggested mechanism for species transmutation similar to his own, it prompted the Down House naturalist to rush his own work to completion with *On the Origin of Species* in November of 1859. Although details are hazy, Wallace likely wrote it in between bouts of malarial fever on the island of Gilolo; he then finished it in February and sent it from Ternate in March, memorializing this otherwise obscure island in the Moluccas group of the eastern Malay Archipelago. Unbeknownst to Wallace, the Ternate letter was read, along with several hastily gathered papers by Darwin, at a meeting of the Linnean Society on July 1, 1858. It can be justly said that modern evolutionary theory by natural selection was born on that day. This letter was subsequently published in the Society's proceedings.

Interestingly, Alfred Russel Wallace never felt he had been treated unfairly by Darwin. Reflecting on the affair years later, Wallace wrote, "Both Darwin and Dr. Hooker wrote to me in the most kind and courteous manner, informing me of what had been done. Of course I not only approved, but felt that they had given me more honour and credit than I deserved, by putting my sudden intuition—hastily written and immediately sent off for the opinion of Darwin and Lyell—on the same level with the prolonged labours of Darwin, who had reached the same point twenty years before me, and had worked continuously during that long period in order that he might be able to present the theory to the world with such a body of systematized facts and arguments as would almost compel conviction. In a letter, Darwin wrote that he owed much to me and his two friends, adding: 'I almost think that Lyell would have proved right, and that I should never have completed my later work'."

Wallace (as he often did) minimized his contribution in this regard. Even a cursory glance shows this letter to be much more than a "sudden intuition" but rather the discovery of a working field naturalist who had carefully observed nature on two ends of the earth. His comments also mask the very real and profound differences between his and Darwin's respective theories. Although these differences were lost upon the Linnean audience, they become apparent with a careful reading of the Ternate letter compared to Darwin's *Origin*.

The letter reissued here is from Wallace's *Contributions to the Theory of Natural Selection. A Series of Essays* published in 1870. Wallace writes that it represents the unaltered original "except one or two grammatical emendations." This version was chosen for two reasons: first, Wallace's minor "emendations" do add clarity; second, the original included no subheadings, which makes for a laborious read. The inclusion of Wallace's subheadings greatly aids the reader in following the author's argument and thought process.

On the Tendency of Varieties to Depart Indefinitely From the Original Type

Instability of Varieties supposed to prove the permanent distinctness of Species.

One of the strongest arguments which have been adduced to prove the original and permanent distinctness of species is, that *varieties* produced in a state of domesticity are more or less unstable, and often have a tendency, if left to themselves, to return to the normal form of the parent species; and this instability is considered to be a distinctive peculiarity of all varieties, even of those occurring among wild animals in a state of nature, and to constitute a provision for preserving unchanged the originally created distinct species.

In the absence or scarcity of facts and observations as to *varieties* occurring among wild animals, this argument has had great weight with naturalists, and has led to a very general and somewhat prejudiced belief in the stability of species. Equally general, however, is the belief in what are called "permanent or true varieties,"—races of animals which continually propagate their like, but which differ so slightly (although constantly) from some other race, that the one is considered to be a *variety* of the other. Which is the *variety* and which the original *species*, there is generally no means of determining, except in those rare cases in which the one race has been known to produce an offspring unlike itself and resembling the other. This, however, would seem quite incompatible with the "permanent invariability of species," but the difficulty is overcome by assuming that such varieties have strict limits, and can never again vary further from the original type, although they may return to it, which, from the analogy of the domesticated animals, is considered to be highly probable, if not certainly proved.

It will be observed that this argument rests entirely on the assumption, that *varieties* occurring in a state of nature are in all respects analogous to or even identical with those of domestic animals, and are governed by the same laws as regards their permanence or further variation. But it is the object of the present paper to show that this assumption is altogether false, that there is a general principle in nature which will cause many *varieties* to survive the parent species, and to give rise to successive variations departing further and further from the original type, and which also produces, in domesticated animals, the tendency of varieties to return to the parent form.

The Struggle for Existence.

The life of wild animals is a struggle for existence. The full exertion of all their faculties and all their energies is required to preserve their own existence and provide for that of their infant offspring. The possibility of procuring food during the least favourable seasons, and of escaping the attacks of their most dangerous enemies, are the primary conditions

which determine the existence both of individuals and of entire species.
These conditions will also determine the population of a species; and by
a careful consideration of all the circumstances we may be enabled to
comprehend, and in some degree to explain, what at first sight appears
so inexplicable—the excessive abundance of some species, while others
closely allied to them are very rare.

The Law of Population of Species.

The general proportion that must obtain between certain groups of animals is readily seen. Large animals cannot be so abundant as small ones;
the carnivora must be less numerous than the herbivora; eagles and lions
can never be so plentiful as pigeons and antelopes; the wild asses of the
Tartarian deserts cannot equal in numbers the horses of the more luxu-
riant prairies and pampas of America. The greater or less fecundity of an
animal is often considered to be one of the chief causes of its abundance
or scarcity; but a consideration of the facts will show us that it really has
little or nothing to do with the matter. Even the least prolific of animals
would increase rapidly if unchecked, whereas it is evident that the animal population of the globe must be stationary, or perhaps, through the
influence of man, decreasing. Fluctuations there may be; but permanent
increase, except in restricted localities, is almost impossible. For example,
our own observation must convince us that birds do not go on increasing every year in a geometrical ratio, as they would do, were there not
some powerful check to their natural increase. Very few birds produce
less than two young ones each year, while many have six, eight, or ten;
four will certainly be below the average; and if we suppose that each pair
produce young only four times in their life, that will also be below the
average, supposing them not to die either by violence or want of food.
Yet at this rate how tremendous would be the increase in a few years
from a single pair! A simple calculation will show that in fifteen years
each pair of birds would have increased to nearly ten millions![1] whereas
we have no reason to believe that the number of the birds of any country

---

1. This is under estimated. The number would really amount to more than two thousand
   millions!

increases at all in fifteen or in one hundred and fifty years. With such powers of increase the population must have reached its limits, and have become stationary, in a very few years after the origin of each species. It is evident, therefore, that each year an immense number of birds must perish—as many in fact as are born; and as on the lowest calculation the progeny are each year twice as numerous as their parents, it follows that, whatever be the average number of individuals existing in any given country, *twice that number must perish annually,*—a striking result, but one which seems at least highly probable, and is perhaps under rather than over the truth. It would therefore appear that, as far as the continuance of the species and the keeping up the average number of individuals are concerned, large broods are superfluous. On the average all above one become food for hawks and kites, wild cats and weasels, or perish of cold and hunger as winter comes on. This is strikingly proved by the case of particular species; for we find that their abundance in individuals bears no relation whatever to their fertility in producing offspring.

Perhaps the most remarkable instance of an immense bird population is that of the passenger pigeon of the United States, which lays only one, or at most two eggs, and is said to rear generally but one young one. Why is this bird so extraordinarily abundant, while others producing two or three times as many young are much less plentiful? The explanation is not difficult. The food most congenial to this species, and on which it thrives best, is abundantly distributed over a very extensive region, offering such differences of soil and climate, that in one part or another of the area the supply never fails. The bird is capable of a very rapid and long-continued flight, so that it can pass without fatigue over the whole of the district it inhabits, and as soon as the supply of food begins to fail in one place is able to discover a fresh feeding-ground. This example strikingly shows us that the procuring a constant supply of wholesome food is almost the sole condition requisite for ensuring the rapid increase of a given species, since neither the limited fecundity, nor the unrestrained attacks of birds of prey and of man are here sufficient to check it. In no other birds are these peculiar circumstances so strikingly combined. Either their food is more liable to failure, or they have not sufficient power of wing to search for it over an extensive area, or during some season of the year it becomes very scarce, and less wholesome substitutes have to be found; and thus, though more fertile in offspring, they can never increase beyond the supply of food in the least favourable seasons.

Many birds can only exist by migrating, when their food becomes scarce, to regions possessing a milder, or at least a different climate, though, as these migrating birds are seldom excessively abundant, it is evident that the countries they visit are still deficient in a constant and abundant supply of wholesome food. Those whose organization does not permit them to migrate when their food becomes periodically scarce, can never attain a large population. This is probably the reason why woodpeckers are scarce with us, while in the tropics they are among the most abundant of solitary birds. Thus the house sparrow is more abundant than the redbreast, because its food is more constant and plentiful,—seeds of grasses being preserved during the winter, and our farm-yards and stubble-fields furnishing an almost inexhaustible supply. Why, as a general rule, are aquatic, and especially sea birds, very numerous in individuals? Not because they are more prolific than others, generally the contrary; but because their food never fails, the sea-shores and river-banks daily swarming with a fresh supply of small mollusca and crustacea. Exactly the same laws will apply to mammals. Wild cats are prolific and have few enemies; why then are they never as abundant as rabbits? The only intelligible answer is, that their supply of food is more precarious. It appears evident, therefore, that so long as a country remains physically unchanged, the numbers of its animal population cannot materially increase. If one species does so, some others requiring the same kind of food must diminish in proportion. The numbers that die annually must be immense; and as the individual existence of each animal depends upon itself, those that die must be the weakest—the very young, the aged, and the diseased,—while those that prolong their existence can only be the most perfect in health and vigour—those who are best able to obtain food regularly, and avoid their numerous enemies. It is, as we commenced by remarking, "a struggle for existence," in which the weakest and least perfectly organized must always succumb.

The Abundance or Rarity of a Species dependent upon its more or less perfect Adaptation to the Conditions of Existence.

It seems evident that what takes place among the individuals of a species must also occur among the several allied species of a group,—viz. that those which are best adapted to obtain a regular supply of food, and to defend themselves against the attacks of their enemies and the vicissitudes of the seasons, must necessarily obtain and preserve a superiority in population; while those species which from some defect of power or organization are the least capable of counteracting the vicissitudes of food, supply, &c., must diminish in numbers, and, in extreme cases, become altogether extinct. Between these extremes the species will present various degrees of capacity for ensuring the means of preserving life; and it is thus we account for the abundance or rarity of species. Our ignorance will generally prevent us from accurately tracing the effects to their causes; but could we become perfectly acquainted with the organization and habits of the various species of animals, and could we measure the capacity of each for performing the different acts necessary to its safety and existence under all the varying circumstances by which it is surrounded, we might be able even to calculate the proportionate abundance of individuals which is the necessary result. If now we have succeeded in establishing these two points—1st, *that the animal population of a country is generally stationary, being kept down by a periodical deficiency of food, and other checks; and, 2nd, that the comparative abundance or scarcity of the individuals of the several species is entirely due to their organization and resulting habits, which, rendering it more difficult to procure a regular supply of food and to provide for their personal safety in some cases than in others, can only be balanced by a difference in the population which have to exist in a given area*—we shall be in a condition to proceed to the consideration of varieties, to which the preceding remarks have a direct and very important application.

Useful Variations will tend to Increase; useless or hurtful Variation to Diminish.

Most or perhaps all the variations from the typical form of a species must have some definite effect, however slight, on the habits or capacities of the individuals. Even a change of colour might, by rendering them more or less distinguishable, affect their safety; a greater or less development of hair might modify their habits. More important changes, such as an increase in the power or dimensions of the limbs or any of the external organs, would more or less affect their mode of procuring food or the range of country which they inhabit. It is also evident that most changes would affect, either favourably or adversely, the powers of prolonging existence. An antelope with shorter or weaker legs must necessarily suffer more from the attacks of the feline carnivora; the passenger pigeon with less powerful wings would sooner or later be affected in its powers of procuring a regular supply of food; and in both cases the result must necessarily be a diminution of the population of the modified species. If, on the other hand, any species should produce a variety having slightly increased powers of preserving existence, that variety must inevitably in time acquire a superiority in numbers. These results must follow as surely as old age, intemperance, or scarcity of food produce an increased mortality. In both cases there may be many individual exceptions; but on the average the rule will invariably be found to hold good. All varieties will therefore fall into two classes—those which under the same conditions would never reach the population of the parent species, and those which would in time obtain and keep a numerical superiority. Now, let some alteration of physical conditions occur in the district—a long period of drought, a destruction of vegetation by locusts, the irruption of some new carnivorous animal seeking "pastures new"—any change in fact tending to render existence more difficult to the species in question, and tasking its utmost powers to avoid complete extermination; it is evident that, of all the individuals composing the species, those forming the least numerous and most feebly organized variety would suffer first, and, were the pressure severe, must soon become extinct. The same causes continuing in action, the parent species would next suffer, would gradually diminish in numbers, and with a recurrence of similar unfavourable conditions might also become extinct. The superior variety would then alone remain, and on a return to favourable circumstances would rapidly increase in numbers and occupy the place of the extinct species and variety.

Superior Varieties will ultimately Extirpate the original Species. The *variety* would now have replaced the *species*, of which it would be a more perfectly developed and more highly organized form. It would be in all respects better adapted to secure its safety, and to prolong its individual existence and that of the race. Such a variety *could not* return to the original form; for that form is an inferior one, and could never compete with it for existence. Granted, therefore, a "tendency" to reproduce the original type of the species, still the variety must ever remain preponderant in numbers, and under adverse physical conditions *again alone survive.* But this new, improved, and populous race might itself, in course of time, give rise to new varieties, exhibiting several diverging modifications of form, any of which, tending to increase the facilities for preserving existence, must, by the same general law, in their turn become predominant. Here, then, we have *progression and continued divergence* deduced from the general laws which regulate the existence of animals in a state of nature, and from the undisputed fact that varieties do frequently occur. It is not, however, contended that this result would be invariable; a change of physical conditions in the district might at times materially modify it, rendering the race which had been the most capable of supporting existence under the former conditions now the least so, and even causing the extinction of the newer and, for a time, superior race, while the old or parent species and its first inferior varieties continued to flourish. Variations in unimportant parts might also occur, having no perceptible effect on the life-preserving powers; and the varieties so furnished might run a course parallel with the parent species, either giving rise to further variations or returning to the former type. All we argue for is, that certain varieties have a tendency to maintain their existence longer than the original species, and this tendency must make itself felt; for though the doctrine of chances or averages can never be trusted to on a limited scale, yet, if applied to high numbers, the results come nearer to what theory demands, and, as we approach to an infinity of examples, become strictly accurate. Now the scale on which nature works is so vast—the numbers of individuals and periods of time with which she deals approach so near to infinity, that any cause, however slight, and however liable to be veiled and counteracted by accidental circumstances, must in the end produce its full legitimate results.

The Partial Reversion of Domesticated Varieties explained.
Let us now turn to domesticated animals, and inquire how varieties produced among them are affected by the principles here enunciated. The essential difference in the condition of wild and domestic animals is this,—that among the former, their well-being and very existence depend upon the full exercise and healthy condition of all their senses and physical powers, whereas, among the latter, these are only partially exercised, and in some cases are absolutely unused. A wild animal has to search, and often to labour, for every mouthful of food—to exercise sight, hearing, and smell in seeking it, and in avoiding dangers, in procuring shelter from the inclemency of the seasons, and in providing for the subsistence and safety of its offspring. There is no muscle of its body that is not called into daily and hourly activity; there is no sense or faculty that is not strengthened by continual exercise. The domestic animal, on the other hand, has food provided for it, is sheltered, and often confined, to guard it against the vicissitudes of the seasons, is carefully secured from the attacks of its natural enemies, and seldom even rears its young without human assistance. Half of its senses and faculties are quite useless;

and the other half are but occasionally called into feeble exercise, while even its muscular system is only irregularly called into action.

Now when a variety of such an animal occurs, having increased power or capacity in any organ or sense, such increase is totally useless, is never called into action, and may even exist without the animal ever becoming aware of it. In the wild animal, on the contrary, all its faculties and powers being brought into full action for the necessities of existence, any increase becomes immediately available, is strengthened by exercise, and must even slightly modify the food, the habits, and the whole economy of the race. It creates as it were a new animal, one of superior powers, and which will necessarily increase in numbers and outlive those inferior to it.

Again, in the domesticated animal all variations have an equal chance of continuance; and those which would decidedly render a wild animal unable to compete with its fellows and continue its existence are no disadvantage whatever in a state of domesticity. Our quickly fattening pigs, short-legged sheep, pouter pigeons, and poodle dogs could never have come into existence in a state of nature, because the very first step towards such inferior forms would have led to the rapid extinction of the race; still less could they now exist in competition with their wild allies. The great speed but slight endurance of the race horse, the unwieldy strength of the ploughman's team, would both be useless in a state of nature. If turned wild on the pampas, such animals would probably soon become extinct, or under favourable circumstances might each lose those extreme qualities which would never be called into action, and in a few generations would revert to a common type, which must be that in which the various powers and faculties are so proportioned to each other as to be best adapted to procure food and secure safety,—that in which by the full exercise of every part of his organization the animal can alone continue to live. Domestic varieties, when turned wild, *must* return to something near the type of the original wild stock, *or become altogether extinct.*[2]

We see, then, that no inferences as to the permanence of varieties in a state of nature can be deduced from the observation of those occurring among domestic animals. The two are so much opposed to each other in every circumstance of their existence, that what applies to the one is almost sure not to apply to the other. Domestic animals are abnormal, irregular, artificial; they are subject to varieties which never occur and never can occur in a state of nature: their very existence depends altogether on human care; so far are many of them removed from that just proportion of faculties, that true balance of organization, by means of which alone an animal left to its own resources can preserve its existence and continue its race.

Lamarck's Hypothesis very different from that now advanced.
The hypothesis of Lamarck—that progressive changes in species have been produced by the attempts of animals to increase the development of their own organs, and thus modify their structure and habits—has been repeatedly and easily refuted by all writers on the subject of varieties and species, and it seems to have been considered that when this was done the whole question has been finally settled; but the view here developed renders such an hypothesis quite unnecessary, by showing that similar results must be produced by the action of principles constantly at work in nature. The powerful retractile talons of the falcon- and the cat-tribes have not been produced or increased by the volition of those animals; but among the different varieties which occurred in the earlier and less highly organized forms of these groups, those always survived longest which had the greatest facilities for seizing their prey. Neither did the giraffe acquire its long neck by desiring to reach the foliage of the more lofty shrubs, and constantly stretching its neck for the purpose, but because any varieties which occurred among its antitypes with

---

2. That is, they will vary, and the variations which tend to adapt them to the wild state, and therefore approximate them to wild animals, will be preserved. Those individuals which do not vary sufficiently will perish.

a longer neck than usual at once secured a fresh range of pasture over the same ground as their shorter-necked companions, and on the first scarcity of food were thereby enabled to outlive them. Even the peculiar colours of many animals, especially insects, so closely resembling the soil or the leaves or the trunks on which they habitually reside, are explained on the same principle; for though in the course of ages varieties of many tints may have occurred, yet those races having colours best adapted to concealment from their enemies would inevitably survive the longest. We have also here an acting cause to account for that balance so often observed in nature,—a deficiency in one set of organs always being compensated by an increased development of some others—powerful wings accompanying weak feet, or great velocity making up for the absence of defensive weapons; for it has been shown that all varieties in which an unbalanced deficiency occurred could not long continue their existence. The action of this principle is exactly like that of the centrifugal governor of the steam engine, which checks and corrects any irregularities almost before they become evident; and in like manner no unbalanced deficiency in the animal kingdom can ever reach any conspicuous magnitude, because it would make itself felt at the very first step, by rendering existence difficult and extinction almost sure soon to follow. An origin such as is here advocated will also agree with the peculiar character of the modifications of form and structure which obtain in organized beings—the many lines of divergence from a central type, the increasing efficiency and power of a particular organ through a succession of allied species, and the remarkable persistence of unimportant parts such as colour, texture of plumage and hair, form of horns or crests, through a series of species differing considerably in more essential characters. It also furnishes us with a reason for that "more specialized structure" which Professor Owen states to be a characteristic of recent compared with extinct forms, and which would evidently be the result of the progressive modification of any organ applied to a special purpose in the animal economy.

Conclusion.

We believe we have now shown that there is a tendency in nature to the continued progression of certain classes of varieties further and further from the original type—a progression to which there appears no reason to assign any definite limits—and that the same principle which produces this result in a state of nature will also explain why domestic varieties have a tendency, when they become wild, to revert to the original type. This progression, by minute steps, in various directions, but always checked and balanced by the necessary conditions, subject to which alone existence can be preserved, may, it is believed, be followed out so as to agree with all the phenomena presented by organized beings, their extinction and succession in past ages, and all the extraordinary modifications of form, instinct and habits which they exhibit.

B.

DARWINISM

AN EXPOSITION OF THE THEORY OF NATURAL SELECTION WITH SOME OF ITS APPLICATIONS

Alfred Russel Wallace

INTRODUCTION BY MICHAEL FLANNERY

Wallace first broke with Darwin in a few concluding sentences of an article appearing in the April 1869 issue of the *Quarterly Review*. Calling upon an "Overruling Intelligence" to account for the mind of man, he further elaborated in an essay published a year later titled, "The Limits of Natural Selection as Applied to Man." The following excerpt reprinted here is taken from chapter 15 ("Darwinism Applied to Man") of his book, *Darwinism* (1889). It represents a further expansion upon his ideas of design and purpose in the natural world. The publication of *Darwinism* led to a nasty fray with George John Romanes, who, not without some justification, called the work a misleading explication of "pure Wallaceism."

Wallace's opening argument here is an intricate one and requires some explanation. Acknowledging that these mental faculties were demonstrable in the "savage" might imply an inheritance from primordial ancestry precisely as Darwin argued. Instead, Wallace believed these faculties to be latent. Elsewhere Wallace had written, "On the whole... we may conclude that the general, moral, and intellectual development of the savage is not less removed from that of civilised man... and from the fact that all moral and intellectual faculties do occasionally manifest themselves, we may fairly conclude that they are always latent, and that the large brain of the savage man is much beyond his actual requirements in the savage state." (See Wallace's *Natural Selection and Tropical Na-*

ture [London: Macmillan, 1895], p. 192.) He even extended this to certain physical attributes in man. "[H]ow can we conceive that early man, as an animal, gained anything by purely erect locomotion? Again, the hand of man contains latent capacities and powers which are unused by savages, and must have been even less used by palæolithic man and his still ruder predecessors. It has all the appearance of an organ prepared for the use of civilized man, and one which was required to render civilization possible" (p. 198). The teleology implied in "prepared" is unmistakable.

The Interpretation of the Facts

THE FACTS now set forth prove the existence of a number of mental faculties which either do not exist at all or exist in a very rudimentary condition in savages, but appear almost suddenly and in perfect development in the higher civilised races. These same faculties are further characterised by their sporadic character, being well developed only in a very small proportion of the community; and by the enormous amount of variation in their development, the higher manifestations of them being many times — perhaps a hundred or a thousand times — stronger than the lower. Each of these characteristics is totally inconsistent with any action of the law of natural selection in the production of the faculties referred to; and the facts, taken in their entirety, compel us to recognise some origin for them wholly distinct from that which has served to account for the animal characteristics — whether bodily or mental — of man.

The special faculties we have been discussing clearly point to the existence in man of something which he has not derived from his animal progenitors — something which we may best refer to as being a spiritual essence or nature, capable of progressive development under favorable conditions. On the hypothesis of this spiritual nature, superadded to the animal nature of man, we are able to understand much that is otherwise mysterious or unintelligible in regard to him, especially the enormous influence of ideas, principles, and beliefs over his whole life and actions. Thus alone we can understand the constancy of the martyr, the unselfishness of the philanthropist, the devotion of the patriot, the enthusiasm of the artist, and the resolute and preserving search of the scientific worker after nature's secrets. Thus we may perceive that the love of truth, the delight in beauty, the passion for justice, and the thrill of exultation with which we hear of any act of courageous self-sacrifice, are the workings within us of a higher nature which has not been developed by means of the struggle for material existence.

It will, no doubt, be urged that the admitted continuity of man's progress from the brute does not admit of the introduction of new causes, and that we have no evidence of the sudden change in nature which such introduction would bring about. The fallacy as to new causes involving any breach of continuity, or any sudden or abrupt change, in the effects, has already been shown; but we will further point out that there are at least three stages in the development of the organic world when some new cause or power must necessarily have come into action.

The first stage is the change from inorganic to organic, when the earliest vegetable cell, or the living protoplasm out of which it arose, first appeared. This is often imputed to a mere increase of complexity of chemical compounds; but increase of complexity, with consequent instability, even if we admit that it may have produced protoplasm as a chemical compound, could certainly not have produced any *living* protoplasm — protoplasm which has the power of growth and of reproduction, and of that continuous process of development which has resulted in the marvellous variety and complex organisation of the whole vegetable kingdom. There is in all this something quite beyond and apart from chemical changes, however complex; and it has been well said that the first vegetable cell was a new thing in the world, possessing altogether new powers — that of extracting and fixing carbon from carbon dioxide of the atmosphere, that of indefinite reproduction, and, still more marvellous, the power of variation and of reproducing those variations till endless complications of structure and varieties of form have been the result. Here, then, we have indications of a new power at work, which we may term *vitality*, since it gives to certain forms of matter all those characters and properties which constitute life.

The next stage is still more marvellous, still more completely beyond all possibility of explanation by matter, its laws and forces. It is the introduction of sensation or consciousness, constituting the fundamental distinction between animal and vegetable kingdoms. Here all idea of mere complication of structure producing the result is out of the question. We feel it altogether preposterous to assume that at a certain stage of complexity of atomic constitution, and as a necessary result of that complexity alone, an *ego* should start into existence, a thing that *feels*, that is *conscious* of its own existence. Here we have the certainty that something new has arisen, a being whose nascent consciousness has gone on increasing in power and definiteness till it has culminated in the higher animals. No verbal explanation or attempt at explanation — such as the statement that life is the result of molecular forces of the protoplasm, or that the whole existing organic universe from the amœba up to man was latent in the fire-mist from which the solar system was developed — can afford any mental satisfaction, or help us in any way to a solution of the mystery.

The third stage is, as we have seen, the existence in man of a number of his most characteristic and noblest faculties, those which raise him furthest above the brutes and open up possibilities of almost infinite advancement. These faculties could not possibly have been developed by means of the same laws which have determined the progressive development of the organic world in general, and also of man's physical organism. (For an earlier discussion of this subject, with some wider applications, see the author's *Contributions to the Theory of Natural Selection*, chap. x.)

These three stages of progress from the inorganic world of matter and motion up to man, point clearly to an unseen universe — to a world of spirit, to which the world of matter is altogether subordinate. To this spiritual world we may refer the marvellously complex forces which we know as gravitation, cohesion, chemical force, radiant force, and electricity, without which the material universe could not exist for a moment in its present form, and perhaps not at all, since without these forces, and perhaps others which may be termed atomic, it is doubtful whether matter itself could have any existence. And still more surely can we refer to it those progressive manifestations of Life in the vegetable, the animal, and man — which we may classify as unconscious, conscious, and intellectual life — and which probably depend upon different degrees of spiritual influx. I have already shown that this involves no necessary infraction of the law of continuity in physical or mental evolution; whence it follows that any difficulty we may find in discriminating the inorganic from the organic, the lower vegetable from the lower animal organisms, or the higher animals from the lowest types of man, has no bearing at all upon the question. This is to be decided by showing that a change in essential nature (due, probably, to causes of a higher order than those of the material universe) took place at the several stages of progress I have indicated; a change which may be none the less real because absolutely imperceptible at its point of origin, as is the change that takes place in the curve in which a body is moving when the application of some new force causes the curve to be slightly altered.

Concluding Remarks

THOSE WHO admit my interpretation of the evidence now adduced — strictly scientific evidence in its appeal to facts which are clearly what ought *not* to be on the materialistic theory — will be able to accept the spiritual nature of man, as not in any way inconsistent with the theory of evolution, but as dependent on those fundamental laws and causes which furnish the very material for evolution to work with. They will also be releaved from the crushing mental burthen imposed upon those who — maintaining that we, in common with the rest of nature, are but products of the blind eternal forces of the universe, and believing also that the time must come when the sun will lose his heat and all life on the earth necessarily cease — have to contemplate a not very distant future in which all this glorious earth — which for untold millions of years has been slowly developing forms of life and beauty to culminate at last in man — shall be as if it never existed; who are compelled to suppose that all the slow growths of our race struggling towards a higher life, all the agony of martyrs, all the groans of victims, all the evil and misery and undeserved suffering of the ages, all the struggles for freedom, all the efforts towards justice, all the aspirations for virtue and the wellbeing of humanity, shall absolutely vanish, and, "like the baseless fabric of a vision, leave not a wrack behind."

As contrasted with this hopeless and soul-deadening belief, we, who accept the existence of a spiritual world, can look upon the universe as a grand consistent whole adapted in all its parts to the development of spiritual beings capable of indefinite life and perfectibility. To us, the whole purpose, the only *raison d'être* of the world—with all its complexities of physical structure, with its grand geological progress, the slow evolution of the vegetable and animal kingdoms, and the ultimate appearance of man — was the development of the human spirit in association with the human body. From the fact that the spirit of man — the man himself — *is* so developed, we may well believe that this is the only, or at least the best, way for its development; and we may even see in what is usually termed "evil" on the earth, one of the most efficient means of its growth. For we know that the noblest faculties of man are strengthened and perfected by struggle and effort; it is by unceasing warfare against physical evils and in the midst of difficulty and danger that energy, courage, self-reliance, and industry have become the common qualities of the northern races; it is by the battle with moral evil in all its hydra-headed forms, that the still nobler qualities of justice and mercy and humility and self-sacrifice have been steadily increased in the world. Being thus trained and strengthened by their surroundings, and possessing latent faculties capable of such noble development, are surely destined for a higher and more permanent existence; and we may confidently believe with our greatest living poet–

> That life is not as idle ore,
> But iron dug from central gloom,

And heated hot with burning fears
And dipt in baths of hissing tears,
And batter'd with the shocks of doom
To shape and use.

We thus find that the Darwinian theory, even when carried out to its extreme logical conclusion, not only does not oppose, but lends a decided support to, a belief in the spiritual nature of man. It shows us how man's body may have been developed from that of a lower animal form under the law of natural selection; but it also teaches us that we possess intellectual and moral faculties which could not have been so developed, but must have had another origin; and for this origin we can only find an adequate cause in the unseen universe of Spirit.

C.

NEW THOUGHTS ON EVOLUTION

BEING THE VIEWS OF DR. ALFRED RUSSEL WALLACE, O.M., F.R.S.
AS GATHERED IN AN INTERVIEW BY HAROLD BEGBIE. REPRINTED FROM
"THE DAILY CHRONICLE" BY KIND PERMISSION OF AUTHOR AND EDITOR,
AND NOW SET FORTH BY CHAPMAN AND HALL, LTD., LONDON [1910]

by Harold Begbie with Alfred Russel Wallace

Introduction by Michael Flannery

This interview presents one of the clearest representations of Wallace's mature thought on the origin of life, evolutionary biology, and cosmology. Replete with suggestions of a designed and guided universe, Wallace's exchange with Begbie is at once a bold declaration for intelligent evolution and a scathing indictment of materialism.

I.

ON THE beautiful and lonely road between Poole Harbour and Broadstone, a matter of three miles, I passed at different times eight or nine tramps the sorriest and most depressing specimens of the human race imaginable; some men, some women, two of them the merest boys.

At the end of my journey I found Professor Alfred Russel Wallace in his study, surrounded by all the pleasant signs of a scholar's ceaseless activity. He is 87 years of age. His eyes shine with intelligence, his movements are quick and active, there is vigour, force, and power in his voice. Tall and spare, with a face of ivory and hair as white as snow, this greatest living representative of the Victorians, the friend and contemporary of Darwin, and with Darwin the simultaneous formulator of the evolution hypothesis, advertises to the world, at the age of 87, the blessings of work and the satisfaction of unresting aspiration. To give has been the gospel of his life—to give himself to the pursuit of truth and to give to mankind the full harvest of his toil. And the result is an old age overflowing with happiness, an intelligence extraordinarily acute, senses unimpaired, and so wide and catholic a delight in human life that he is able to sympathise with the youngest of our dreams and to follow the political progress of humanity with enthusiasm only bridled by amusement at our old-fashioned delays and hesitancies.

It is perhaps necessary to say at the beginning of this article that there are people in the world who maintain that the hypothesis of evolution has explained everything, that the universe is self-contained and self-sufficient, that the law called Uniformity of Nature makes a controlling God unthinkable and impossible, that there has never been a creation, that there has never been a scheme, and that there is no purpose in anything. All is accident, chance, and meaningless haphazard. "The world is a condensation of primeval gas, a congeries of stones and meteors." These people are not Agnostics; Agnostics only say that they do not know how and why things have come to pass; and they are not Monists; no, they are Materialists, the fighting force of a polite and hesitating scepticism, the definite challengers and onslaughterers of anything in the nature of Idealism; and instead of saying that they do not know this or that, they claim very emphatically, with gallant Prince Haeckel at their head, to know all. And their all is nothing.

But how did life begin upon this planet?

In its home this earth of ours was part of the sun. Now, the sun is very hot. A kettle of boiling water or a cauldron of liquid lead would be icy even to the outmost fringe of its enveloping flames. We have no idea how hot is the heart of the sun, out of which our earth leaped some millions of aeons ago, and set up as a colony on its own account, only attached to the mother-land by the sentimental tie of gravitation. For no one knows how many cycles this emigrant earth was a flaming and roaring ball of conflagration, then for thousands of years it was hotter than anything we can imagine, and when it settled down into its stride it was about as habitable as the lava of Mount Vesuvius. More millions of years (as many as you please) and the surface of the earth cooled, cooled at last so wonderfully that plants were able to grow, creatures to appear, and finally we get Esquimaux shivering in furs and the indomitable Dr. Cook taking off his hat to the North Pole.

Now, I have never been able to understand how the germ of life managed to exist in our molten earth. How did it endure that unthinkable heat? If I drop seeds into the fire, or boil them in a kettle, they either disappear or refuse to grow. But they were once part and parcel of a blazing furnace, they were contained in a condensation of primeval gas, they existed in the sun. Miracle of miracles! The Agnostic tells me he does not know. The Materialist says it is only a little more difficult than that ancient problem, Which came first, the hen or the egg? But I do not know the answer to that highly important riddle, and the Materialist does not tell me anything I can comprehend on the matter. All my ignorance on this subject I laid humbly at the feet of Professor Wallace, Father of Evolution and most open-minded of observers.

"Of course," he said, with a smile, "there is no reasonable answer possible to Materialism. Life could not have existed on the red-hot planet. No life at all, not the lowest and obscurest forms. Materialists know this. Some of them get out of the difficulty by saying that life was rained upon the earth in meteors! That is a theory more amusing than ridiculous. We need not discuss it."

"But what is the answer?"

"Well, it is the very simple, plain, and old-fashioned one, that there was at some stage in the history of tile earth, after the cooling process, a definite act of creation. Something came from the outside. Power was exercised from without. In a word, life was given to the earth. All the errors of those who have distorted the thesis of evolution into something called, inappropriately enough, Darwinism, have arisen from the supposition that life is a consequence of organisation. This is unthinkable. Life, as Huxley admitted, is the cause and not the consequence of organisation. Admit life, and the hypothesis of evolution is sufficient and unanswerable. Postulate organisation first, and make it the origin and cause of life, and you lose yourself in a maze of madness. An honest and unswerving scrutiny of nature forces upon the mind this certain truth, that at some period of the earth's history there was an act of creation, a giving to the earth of something which before it had not possessed; and from that gift, the gift of life, has come the infinite and wonderful population of living forms. Then, as you know, I hold that there was a subsequent act of creation, a giving to man, when he had emerged from his ape-like ancestry, of a spirit or soul. Nothing in evolution can account for the soul of man. The difference between man and the other animals is unbridgeable. Mathematics is alone sufficient to prove in man the possession of a faculty unexistent in other creatures. Then you have music and the artistic faculty. No, the soul was a separate creation."

"But are these the only two instances of interference from outside?"

"Ah, we come to a great question. I deal with it in a book which Chapman and Hall are to publish this winter. In some ways this book will be my final contribution to the philosophic side of evolution. It concerns itself with the great question of Purpose. Is there guidance and control, or is everything the result of chance? Are we solitary in the cosmos, and without meaning to the rest of the universe; or are we one in 'a stair of creatures,' a hierarchy of beings? Now, you may approach this matter along the metaphysical path, or, as a man of exact science, by observation of the physical globe and reflection upon visible and tangible objects. My contribution is made as a man of science, as a naturalist, as a man who studies his surroundings to see where he is. And the conclusion I reach in my book is this: That everywhere, not here and there, but everywhere, and in the very smallest operations of nature to which human observation has penetrated, there is Purpose and a continual Guidance and Control."

II.

IT WOULD not be right for me, and it might be dangerous, to attempt anything in the nature of a summary of Professor Wallace's argument as it will appear in book form; but exercising great care and with as good a memory as I can command, I will here mention just one or two instances as quoted by the author, in the ease and carelessness of unstudied conversation, to justify his thesis that there is a continual guidance and control all through this mystery of terrestrial existence. The reader, I hope, will be just enough to form in his mind rather a determination to read the book when it appears than to pass any judgment on the author's argument in this place.

"There seems to me," said Professor Wallace, "unmistakable evidence of guidance and control in the physical apparatus of every living creature. Consider for a moment the question of nourishment. Men of various races eat different foods; men of the same race may follow diets as separate and distinct as chalk from cheese. But in all cases the main result is the same. The food is converted into blood. That is interesting enough, marvelous enough, baffling enough; but mark what follows. This blood circulating through the body becomes at one point hair and at another nail; here it transforms itself into bone and there into tissue; at the same moment that it changes into skin it changes into nerve; it is at once the bone in my finger and the eye in my head. Materialism forges such words as secretion, but no word signifying unconscious and accidental action can explain this mystery.

"Just reflect upon it. The blood in our veins becomes at one point a finger-nail; it becomes hard and horn-like substance, with a recognisable and distinct surface-texture and character. And it becomes, also, the hair on our heads. How does the same fluid, unconsciously and without intelligence, perform these very diverse and marvellous duties? Remember, this activity of the blood is incessant; it continues to the moment of death. The busiest thing on the earth is this mysterious liquid which we call blood. It is building up the horns and hides of animals, the feathers and beaks of birds, the scales and bones of reptiles, the wings and eyes of insects, the brains of poets and the muscles of workmen. It is digesting the food of all of us, repairing our wasted tissues, restoring our energies, making us and remaking us at every hour of the day. Now, is it not a madness to say that blood can do all these quite marvellous and diverse things of itself, that without consciousness and without direction it flows to a finger-tip and accidentally becomes nail, or mounts to the skull and fortuitously becomes hair? Is it more consonant with reason to say that the blood does its work by itself and without meaning to do it, or that it is intelligently controlled to its purpose by a conscious direction? Which is the saner theory?"

I asked my host if he had formulated any opinion as to the nature and character of the guidance which superintends the management of our bodies.

"I believe it to be," he said, "the guidance of beings superior to us in power and intelligence. Call them spirits, angels, gods, what you will; the name is of no importance. I find this control in the lowest cell; the wonderful activity of cells convinces me that it is guided by intelligence and consciousness. I cannot comprehend how any just and unprejudiced mind, fully aware of this amazing activity, can persuade itself to believe that the whole thing is a blind and unintelligent accident. It may not be possible for us to say how the guidance is exercised, and by exactly what powers; but for those who have eyes to see and minds accustomed to reflect, in the minutest cells, in the blood, in the whole earth, and throughout the stellar universe—our own little universe, as one may call it—there is intelligent and conscious direction; in, a word, there is Mind."

"Myers suggested that our normal consciousness is only a fragment of our total soul, that a greater part of us is at work on the body, managing all the wonderful and complex machinery of the organism, and influencing us without our knowledge."

"Yes, that may or may not be true. But we must enlarge our vision. We must see more beings in the universe than ourselves. I think we have got to recognise that between man and the Ultimate God there is an almost infinite multitude of beings working in the universe at tasks as definite and important as any that we have to perform on the earth. I imagine that the universe is peopled with spirits—that is to say, with intelligent beings, with powers and duties akin to our own, but vaster, infinitely vaster. I think there is a gradual ascent from man upwards and onwards, through an almost endless legion of these beings, to the First Cause, of whom it is impossible for us to speak. Through Him these endless beings act and achieve, but He Himself may have no actual contact with our earth."

"Sometimes this management of our bodies breaks down."

"That is true. I do not mean that the control is absolute or that it is of the nature of interference. The control is evidently bound by laws as absolute and irrefragable as those which govern man and his universe. It is certainly dependent on us in a very large measure for its success. I believe we are influenced, not interfered with, and that the management of our bodies is at least as difficult, for those charged with it, as, let us say, the cultivation of this planet for us."

"But, in any case, you believe that there is purpose in creation?"

"It meets me everywhere I turn. I cannot examine the smallest or the commonest living thing without finding my reason uplifted and amazed by the miracle, by the beauty, the power, and the wisdom of its creation. Have you ever examined the feather of a bird? I almost think a feather is the masterpiece of nature. No man in the world could make such a thing, or anything in the very slightest degree resembling it. Someone has said that a single feather from a heron's wing is composed of over a million parts! The quill is socketed, held together by little contrivances in the nature of hooks and eyes; it is of a material so light that a finger can twist it out of shape, but if it gets pierced or separated by any slight blow it becomes quickly reunited and restored. Watch a bird sailing high above the earth in a gale of wind, and then remind yourself of the lightness of its feathers. And those feathers are airtight and waterproof, the perfectest vesture imaginable!

"Have you ever thought of this, too? The loveliest and most graceful thing on the earth is unquestionably a bird. I suppose everybody has felt that. One cannot watch the flight of the least of birds or study the wonders of their plumage, without feeling a quite intense admiration. They are exquisite creations. Well, all the beauty is in the feathers. Strip a bird of its plumage, and what was the perfectest thing becomes at once the most ugly and comical. A young bird makes us laugh. When its feathers have grown, the same bird makes Shelley write an immortal ode. Such is the wonder of feathers. And how do they grow? Evolution can explain a great deal; but the origin of a feather, and its growth, this is beyond our comprehension, certainly beyond the power of accident to achieve."

He shook his head and smiled amiably upon the hot-headedness of Darwinians. "The scales on the wing of a moth," he said, quietly, "have no explanation in evolution. They belong to Beauty, arid Beauty is a spiritual mystery. Even Huxley was puzzled by the beauty of his environment. What is the origin of Beauty? Evolution cannot explain."

"Nevertheless, of course, evolution is a sound hypothesis?"

"Every fresh discovery in nature fortifies that original hypothesis. But this is the sane and honest evolution, which does not concern itself at all with beginnings, and merely follows a few links in a fairly obvious chain. As for the chain itself, evolution has nothing to say. For my own part, I am convinced that at one period in the earth's history there was a definite act of creation, that from that moment evolution has been at work, guidance has been exercised. The more deeply men reflect upon what they are able to observe, the more they will be brought to see that Materialism is a most gigantic foolishness. And I think it will soon pass from the mind. At first there was some excuse. Into the authoritative nonsense and superstitions of Clericalism, evolution threw a bomb of the most deadly power. Those whose intelligence had been outraged and irritated by this absurd priestcraft rushed to the conclusion that religion was destroyed, that a little chain of reasoning had explained the whole infinite universe, that in mud was the origin of mind, and in dust its end. That was an opinion which could not last. Materialism is as dead as priestcraft for all intelligent minds. There are laws of nature, but they are purposeful. Everywhere we look we are confronted by power and intelligence. The future will be full of wonder reverence, and a calm faith worthy of our place in the scheme of things."

"And greater knowledge?"

"Oh, yes, we are only at the beginning of the puzzle."

Endnotes

1. John Wilson, *The Forgotten Naturalist: Alfred Russel Wallace* (Melbourne: Australian Scholarly Publishing, 2000).

2. Sahotra Sarkar, "Wallace's belated revival," *Journal of Bioscience* 23.1 (March 1998): 2–7, 5.

3. See Peter Raby, *Alfred Russel Wallace: A Life* (London: Chatto & Windus, 2001); Michael Shermer, *In Darwin's Shadow: The Life and Science of Alfred Russel Wallace* (New York: Oxford University Press, 2002); Rosse A. Slotten, *The Heretic in Darwin's Court: The Life of Alfred Russel Wallace* (New York: Columbia University Press, 2004); Martin Fichman, *An Elusive Victorian: The Evolution of Alfred Russel Wallace*: Chicago: The University of Chicago Press, 2004); William Bryant, *The Birds of Paradise: Alfred Russel Wallace, A Life* (New York: iUniverse, 2006); and Charles H. Smith and George Beccaloni, eds., *Natural Selection and Beyond: The Intellectual Legacy of Alfred Russel Wallace* (New York: Oxford University Press, 2008).

4. Alfred Russel Wallace, *My Life: A Record of Events and Opinions* (reprint, 2005; London: Chapman & Hall, 1908), p. 5.

5. Ibid., p. 6.

6. Iain McCalman, *Darwin's Armada: Four Voyages and the Battle for the Theory of Evolution* (New York: W. W. Norton, 2009), p. 222.

7. Wallace, *My Life*, p. 28.

8. Ross A. Slotten, p. 11. At his older brother John's urging, Alfred began attending the London Mechanic's Institute, a working-class educational facility. But, as young Alfred soon discovered, such establishments were also often hotbeds of social, political, and philosophical radicalism.

9. Wallace, *My Life*, p. 46.

10. Wallace, *My Life*, pp. 123–124.

11. Charles Darwin, *The Autobiography of Charles Darwin*, edited by Francis Darwin (1893; reprint, Amherst, NY: Prometheus Books, 2000), p. 43.

12. R. Elwyn Hughes, "Alfred Russel Wallace: Some Notes on the Welsh Connection," *The British Society for the History of Science* 22.4 (1989): 401–418.

13. Quoted in Ibid., 403.

14. Ibid., 404.

15. Ibid., 408.

16. Ibid., 409–410.

17. Wallace, *My Life*, pp. 84–85.

18. Ibid., p. 134,

19. Hughes, 413–415.

20. Wallace, *My Life*, p. 144.

21. Slotten gives additional practical reasons for the choice of Pará in his biography, pp. 35–39.

22. John Hemming, "Scholars and Amazon," *Geological* 80.5 (May 2008): 50–54.

23. Slotten, p. 42.

24. Ibid., p. 43.

25. Fichman gives an excellent account of the critical relationship between Wallace and Stevens. See, p. 22.

26. Alfred Russel Wallace, *Travels on the Amazon and Rio Negro: With an Account of the Native Tribes*, new ed. [first published 1853] (London: Ward, Lock, & Co. New ed., 1889), p. 143.

27. Ibid., pp. 334–335.

28. Ibid., p. 344.

29. Ibid., p. 361.

30. Charles Darwin, *Charles Darwin's Voyage of the Beagle Round the World* (New York: Tess Press, [2000?]), pp. 199–200.

[A reprint of *The Voyage of the Beagle*, 2nd ed. London: John Murray, 1845.]

31. McCalman, pp. 239–240.

32. Wallace, *My Life*, p. 159.

33. Slotten, p. 91.

34. See McCalman, p. 251; and Slotten, p. 96 respectively.

35. Fichman, p. 28.

36. Wallace, *My Life*, p. 169.

37. Ibid., p. 174.

38. Ibid., pp. 175–176.

39. Slotten, pp. 109–111.

40. Wallace, *My Life*, p. 178.

41. Alfred Russel Wallace, *The Malay Archipelago: The Land of the Orang-utan and the Bird of Paradise a Narrative of Travel With Studies of Man and Nature* (New York: Harper & Brothers, 1869), p. 54.

42. Ibid., p. 46.

43. Henry Fairfield Osborn, *From the Greeks to Darwin: The Development of the Evolution Idea Through Twenty-Four Centuries* (New York: Charles Scribner's Sons, 1929), pp. 322–323.

44. McCalman, p. 266.

45. Qtd. in Ibid.

46. Slotten, pp. 121–122.

47. Janet Browne, *Charles Darwin: Voyaging* (Princeton: Princeton University Press, 1995), p. 538.

48. Qtd. in Wallace, *My Life*, 183.

49. Ibid., p. 186.

50. Wallace, *The Malay Archipelago*, p. 461.

51. Ibid., p. 486.

52. Ibid., p. 312.

53. Wallace, *My Life*, p. 190.

54. Qtd. in Slotten, p. 147. The precise date of posting is generally acknowledged, see: McCalman, p. 290; Wallace, Alfred Russel, *Dictionary of Scientific Biography*, s.v., H. Lewis McKinney, 1981; Roy Davies, *The Darwin Conspiracy: Origins of a Scientific Crime* (London: Goldensquare Books, 2009), p. 146.

55. Slotten, p. 153.

56. Benjamin Wiker, *The Darwin Myth: The Life and Lies of Charles Darwin* (Washington, DC: Regnery Publishing, 2009), p. 60.

57. This question has plagued historians. For a general review and at least one opinion, see Robert J. Richards, "Why Darwin Delayed, or Interesting Problems and Models in the History of Science," *Journal of the History of the Behavioral Sciences* 19 (January 1983): 45–53.

58. Darwin, *Autobiography*, pp. 180–181.

59. Browne, pp. 516–518.

60. Qtd. in Ibid., p. 514.

61. Slotten, p. 151.

62. Qtd. in Adrian Desmond and James Moore, *Darwin: The Life of a Tormented Evolutionist* (New York: W. W. Norton, 1991), p. 369

63. For details of Darwin's activities with the Plinian Society, see J. H. Ashworth, "Charles Darwin as a Student in Edinburgh, 1825–1827," *Proceedings of the royal Society of Edinburgh* 55 (1935): 97–113.

64. Quoted and discussed at length in Adrian Desmond and James Moore, *Darwin*, p. 32.

65. The incident is discussed more fully in Howard E. Gruber, *Darwin on Man: A Psychological Study of Scientific Creativity together with Darwin's Early and Unpublished Notebooks*, transcribed and annotated by Paul H. Barrett (London: Wildwood House, 1974), p. 39. The text of the expunged minutes has been reprinted in the appendix, p. 479. The upshot is, "That mind as far as one individual sense, & consciousness are concerned, is material."

66. Desmond and Moore, p. 34.

67. Gruber, p. 122

68. Ibid.

69. Stanley L. Jaki, *The Savior of Science* (Washington, DC: Regnery Gateway, 1988), p.126.

70. Silvan S. Schweber, "The Origin of the 'Origin' Revisited," *Journal of the History of Biology* 10.2 (Autumn 1977): 229–316, 233–234.

71. This can be readily deduced from Darwin's Notebook C written from February 1838 through July of that year. For example, he declares, "love of the deity effect of organization. Oh you materialist!" And again shortly follows with a rhetorical question, "Why is thought, being a secretion of brain, more wonderful than gravity a property of matter? It is our arrogance, it is our admiration of ourselves...." See Charles Darwin, *Charles Darwin's Notebooks, 1836–1844*, transcribers and editors Paul H. Barrett, Peter J. Gautrey, Sandra Herbert, et al. (Cambridge, England: Cambridge University Press; London: British Museum, 1987), p. 291.

72. Gillian Beer, *Darwin's Plots: Evolutionary Narrative in Darwin, George Eliot, and Nineteenth-century Fiction*, 2nd ed. (New York: Cambridge University Press, 2000), p. xviii.

73. This is fully explicated in John Angus Campbell, "Why Was Darwin Believed?: Darwin's *Origin* and the Problem of Intellectual Revolution," *Configurations* 11 (2004): 203–237.

74. For a complete account, see Ralph Colp Jr., "'Confessing a Murder': Darwin's First Revelations about Transmutation," *Isis* 77.1 (March 1986): 8–32.

75. Ibid., 21.

76. Ibid., 32.

77. Charles Darwin, *The Life and Letters of Charles Darwin*, edited by Francis Darwin (New York: D. Appleton, 1897), vol. 1, 507.

78. The details of the unveiling of Wallace's and Darwin's theories at the Linnean Society are ably chronicled in Janet Browne, *Charles Darwin: The Power of Place* (Princeton: Princeton University Press, 2002), pp. 33–37

79. Desmond and Moore, p. 467.

80. Browne, *The Power of Place*, p. 36.

81. Ibid., p. 40.

82. Qtd in Ibid., p. 42.

83. Fichman, p. 103; Raby, p. 139; and Arnold C. Brackman, *A Delicate Arrangement: The Strange Case of Charles Darwin and Alfred Russel Wallace* (New York: Times Books, 1980).

84. Gertrude Himmelfarb, *Darwin and the Darwinian Revolution*, rev. ed. (1962; reprint, Chicago: Elephant Paperbacks, 1996), pp. 334–335.

85. Letter from Darwin to Wallace, January 25, 1859, reproduced in James Marchant, *Alfred Russel Wallace: Letters and Reminiscences* (1916; reprint, [n.p.]: Bibliobazaar, 2008), v. 1, 144.

86. See, for example, Barbara Beddall, *Wallace and Bates in the Tropics: An Introduction to the Theory of Natural Selection* (New York: Macmillan, 1969); H. L. McKinney, *Wallace and Natural Selection* (New Haven: Yale University Press, 1972); Arnold C. Brackman, *A Delicate Arrangement* (see note 83 above); and most recently Roy Davies (n. 54).

87. Wallace, *My Life*, p. 193.

88. Qtd. in Slotten, p. 161.

89. Himmelfarb, p. 252.

90. Browne, *The Power of Place*, pp. 88–90.

91. Ibid., pp. 102–103.

92. Wallace, *The Malay Archipelago*, p. 363.

93. For a thorough discussion of biology and the Wallace Line, see Penny van Oosterzee, *Where Worlds Collide: The Wallace Line* (Ithaca: Cornell University Press, 1997), esp. pp. 21–36.

94. Wallace, *The Malay Archipelago*, p. 284.

95. Qtd. in van Oosterzee, p. 34.

96. Alfred Russel Wallace, "On the Zoological Geography of the Malay Archipelago," *Proceedings of the Linnean Society* 156 (1859): 172–183, 174.

97. For details, see Martin Fichman, "Zoogeography and the Problem of Land Bridges," *Journal of the History of Biology* 10.1 (Spring 1977): 45–63.

98. van Oosterzee, pp. 34, 36.

99. Wallace, *The Malay Archipelago*, p. 539.

100. Wallace, *The Malay Archipelago*, p. viii.

101. Raby, p. 162.

102. Ibid., pp. 163–165.

103. Slotten, p. 197.

104. William Cecil Dampier, *A History of Science and Its Relations with Philosophy & Religion*, 3rd ed. (New York: Macmillan, 1946), pp. 274–275.

105. Alfred Russel Wallace, "The Origin of Human Races and the Antiquity of Man Deduced from the Theory of Natural Selection," *Journal of the Anthropological Society of London* 2 (1864): 158–187, 162.

106. Qtd. in Raby, p. 179.

107. Qtd. in Slotten, p. 215.

108. Charles Lyell, *Geological Evidences of the Antiquity of Man* (Philadelphia: George W. Childs, 1863), p. 506.

109. Peter Raby, "The Literary Legacy of Alfred Russel Wallace," in *Natural Selection and Beyond: The Intellectual Legacy of Alfred Russel Wallace*, edited by Charles H. Smith and George Beccaloni (New Yrok: Oxford University Press, 2008), pp. 223–234.

110. Raby, p. 187.

111. Alfred Russel Wallace, *The Scientific Aspect of the Supernatural: Indicating the Desirableness of an Experimental Enquiry by Men of Science into the Alleged Powers of Clairvoyants and Mediums* (London: F. Farrah, 1866).

112. Ibid., p. 7.

113. Wallace, *My Life*, p. 356.

114. This is the essential thesis of Michael Shermer's *In Darwin's Shadow: The Life and Science of Alfred Russel Wallace* (New York: Oxford University Press, 2002).

115. Peter Lamont, "Spiritualism and a Mid-Victorian Crisis in Evidence," *The History Journal* 47.4 (2004): 897–920.

116. Marchant, v. 1, 242.

117. Alfred Russel Wallace, "Sir Charles Lyell on Geological Climates and the Origin of Species," *Quarterly Review* 126 (April 1869): 359–394.

118. Fichman, *An Elusive Victorian*, p. 80.

119. Jean Gayon, *Darwinism's Struggle for Survival: Heredity and the Hypothesis of Natural Selection*, trans. by Matthew Cobb (New York: Cambridge University Press, 1998), p. 35.

120. Wallace, "Sir Charles Lyell," 394.

121. Marchant, v. 1, 244.

122. Ibid., 252.

123. Qtd in Janet Browne, *The Power of Place*, p. 318.

124. The best comparative analysis of Darwin's and Wallace's theories of natural selection is found in Jean Gayon (see note 119), pp. 19–59.

125. Ibid., p. 37.

126. Ibid., p. 59.

127. Phillip E. Johnson, *Darwin on Trial*, 2nd ed. (Downers Grove: IL: InterVarsity Press, 1993), p. 17.

128. Melinda B. Fagan, "Wallace, Darwin, and the Practice of Natural History," *Journal of the History of Biology* 40 (2007): 601– 635, 616

129. Ibid., 618.

130. Ibid., 625.

131. Ibid., 629.

132. Michael J. Behe, *The Edge of Evolution: The Search for the Limits of Darwinism* (New York: Free Press, 2007), p. 72.

133. Johnson, pp. 15–16.

134. Darwin himself saw his theory as "one long argument." See his *On the Origin of Species by Means of Natural Selection or The Preservation of Favored Races in the Struggle for Life* (1859; reprinted, New York: The Classics of Medicine Library, 1998), p. 459.

135. Charles Darwin, *The Life and Letters of Charles Darwin*, v. 2, 105.

136. Charles Darwin, *Autobiography*, p. 63.

137. Charles Darwin, *The Variation of Animals and Plants Under Domestication* (New York: D Appleton, 1897), v. 1, 6–7.

138. Charles Darwin, *The Life and Letter of Charles Darwin*, v. 2, 202–203.

139. On Darwin's references to Hume, see Paul H. Barrett, Peter J. Gautrey, Sandra Herbert, et. al., transcribers and editors, see *Darwin's Notebooks*,, pp. 321, 325, 545, 559, 591, 592, 596. For more on Hume's influence on Darwin, see William B. Huntley, "David Hume and Charles Darwin," *Journal of the History of Ideas* 33.3 (July–Sept. 1972): 457–470. On Darwin's references to Comte, see *Charles Darwin's Notebooks*, pp. 535, 539, 553, 566, 608; and Silvan S. Schweber, "The Young Darwin," *Journal of the History of Biology* 12.l (Spring 1979): 175–192.

140. C. D. Darlington, *Darwin's Place in History* (Oxford: Basil Blackwell, 1959), p. 60.

141. Fichman, *An Elusive Victorian*, p. 204.

142. Darwin, *On the Origin of Species*, p. 127.

143. Excellent summaries and analyses of the X Club activities are available in Iain McCalman, pp. 354–358; and Edward Caudill, "The Bishop-Eaters: The Publicity Campaign for Darwin and on the Origin of Species," *Journal of the History of Ideas* 55.3 (July 1994): 441–460.

144. Caudill, 443.

145. Qtd. in Ibid., 450.

146. Ibid.

147. McCalman, p. 354.

148. Qtd. in Caudill, 441.

149. Ibid., 452.

150. Ernst Haeckel, *The Riddle of the Universe*, trans. by Joseph McCabe (New York: Harper & Brothers, 1901), pp. 79–80.

151. Wallace, *My Life*, p. 224.

152. Charles Darwin, *The Descent of Man* (1871; reprinted, New York: Barnes & Noble Books, 2004), pp. 99–100.

153. Adrian Desmond and James Moore, *Darwin's Sacred Cause* (Boston: Houghton, Mifflin, 2009), p. xviii.

154. Ibid., p. 370.

155. Ibid., p. 318.

156. Charles Darwin, *The Descent of Man*, p. 125.

157. Wallace, *My Life*, p. 233.

158. Ralph Colp Jr., "'I will certainly do my best': How Darwin Obtained a Civil List Pension for Alfred Russel Wallace," *The History of Science Society* 83.1 (March 1992): 2–26.

159. Marchant, v. 1, 308.

160. For details, see James Moore, "That Evolution Destroyed Darwin's Faith in Christianity—Until He Reconverted on His Deathbed," in *Galileo Goes to Jail, and Other Myths About Science and Religion*, edited by Ronald L. Numbers (Cambridge: Harvard University Press, 2009), 142–151. See also, James Moore, *The Darwin Legend* (Grand Rapids, MI: Baker Books, 1994).

161. For details, see Frank Burch Brown, "The Evolution of Darwin's Theism," *Journal of the History of Biology* 19.1 (Spring 1986): 1–45.

162. Browne, *The Power of Place*, p. 177.

163. Darwin, *Autobiography*, p. 66.

164. Qtd. in Browne, *The Power of Place*, p. 94.

165. Charles Hodge, *What is Darwinism?* (1874; reprinted, New York: Bibiobazaar, 2007), p. 110.

166. Charles Darwin, *The Descent of Man*, p. 79.

167. Maurice Mandelbaum, "Darwin's Religious Views," *Journal of the History of Ideas* 19.3 (June 1958): 363–378.

168. Robert Flint, *Agnosticism*, The Croall Lecture for 1887–88 (London: William Blackwood and Sons, 1903), pp. 50–51.

169. Edward J. Larson, *Evolution: The Remarkable History of a Scientific Theory* (New York: The Modern Library, 2004), p. 69.

170. Wiker, p. 109.

171. Karl W. Giberson, *Saving Darwin: How to Be a Christian and Believe in Evolution* (New York: HarperOne, 2008), pp.19–20. See also, Alberto Kornblihtt, "On Intelligent Design, Cognitive Realism, Vitalism and the Mystery of the Real World," *Life* 54.4/5 (April/May 2007): 235–237, 236.

172. Slotten, p. 411.

173. B. E. Bishop, "Mendel's Opposition to Evolution and to Darwin," *Journal of Heredity* 87.3 (1996): 205–213.

174. Qtd. in Fichman, *An Elusive Victorian*, pp. 106–107. For more on Wallace's American tour, see Martin Fichman's "Alfred Russel Wallace's North American tour: transAtlantic evolutionism," *Endeavour* 25.2 (2001): 74–78.

175. Marchant, v. 2, 57.

176. For details, see Fern Elsdon-Baker, "Spirited Dispute: The Secret Split Between Wallace and Romanes," *Endeavour* 32.2 (2008): 75–78

177. David Quammen, *The Reluctant Mr. Darwin: An Intimate Portrait of Charles Darwin and the Making of His Theory*

of Evolution (New York: W. W. Norton, 2006), p. 231.

178. Wallace, *My Life*, p. 237.

179. Fichman, *An Elusive Victorian*, p. 301.

180. Alfred R. Wallace, *Man's Place in the Universe: A Study of the Results of Scientific Research in Relation to the Unity or Plurality of Worlds*, 3rd ed. (London: George Bell and Son, 1904), p. 206.

181. Ibid., p. 327.

182. Ibid., pp. 335–336.

183. For purposes here the 1914 edition by Chapman and Hall scanned for online access in the "American Libraries" database by Internet Archive is used.

184. Wallace, *The World of Life*, pp. vi–vii.

185. Ibid., p. 197.

186. Ibid., p. 287.

187. Ibid., p. 303.

188. Ibid., p. 316.

189. Ibid., p. 333.

190. See Sander Gliboff, *H. G. Bronn, Ernst Haeckel, and the Origins of German Darwinism: A Study in Translation and Transformation* (Cambridge, MA: MIT Press, 2008), p. 118, 128–130.

191. Darwin, *Autobiography*, pp. 242–243.

192. See his letter to Joseph Hooker in *The Life and Letters of Charles Darwin*, v. 2, 202–203.

193. Wallace, *The World of Life*, p. 337.

194. Ibid., p. 350.

195. Ibid., pp. 374–375.

196. Lewis stated, "a great deal of what appears to be animal suffering need not be suffering in any real sense. It may be we who have invented the 'sufferers' by the 'pathetic fallacy' of reading into the beasts a self for which there is no real evidence." See his *The Problem of Pain* (1940; reprinted, San Francisco: HarpersSanFrancisco, 1996), p. 137.

197. See, for example, Samuel Henry Dickson, *Essays on Life, Sleep, Pain, Etc.* (Philadelphia: Blanchard and Lea, 1851), pp. 91–132.

198. Quammen, p. 117.

199. Alfred Russel Wallace, "A Defense of Modern Spiritualism," pt. 1, and "Spirit-Photographs," *Fortnightly Review* 15 (1874): 630–657, 785–807.

200. Wallace, *The World of Life*, p. 400.

201. Qtd. in Marchant, v. 2, 261.

202. *The Darwin-Wallace Celebration Held on Thurday, 1st July, 1908, by the Linnean Society* (London: The Society, 1908), pp. 10–11.

203. Thomas P. Weber, "Alfred Russel Wallace and the Antivaccination Movement in Victorian England," *Emerging Infectious Diseases* 16.4 (April 2010): 664–668.

204. Ibid., 667.

205. Qtd. in T. D. A. Cockerell, "Recollections of Dr. Alfred Russel Wallace," *Science*, new series 38.990 (December 19, 1913): 871–877, 877.

206. See, for example, Wilma George, *Biologist Philosopher: A Study of the Life of and Writing of Alfred Russel Wallace* (New York: Abelbard Schuman, 1964; and Wallace, Alfred Russel, *Dictionary of Scientific Biography*, s.v., H. Lewis McKinney, 1981.

207. J. M. Mello, *The Mystery of Life, With Special Reference to "The World of Life"* (Warwick: Henry H. Lacy, 1911), pp. 18–19.

208. *The Fundamentals: A Testimony to the Truth*, 2 vols., edited by R. A. Torrey, A. C. Dixon, et al., eds. (1917; reprinted, Grand Rapids, MI: Baker Books, 2003), v. 1, 334–347.

209. Ibid., 346.

210. Don B. DeYoung, *Pioneer Explorers of Intelligent Design* (Winona Lake, IN: BMH Books, 2006), p. 82.

211. Shermer, p. viii.

212. Ibid.,p. 250–270.

213. Steven J. Dick, "The Universe and Alfred Russel Wallace," in *Natural Selection and Beyond: The Intellectual Legacy of Alfred Russel Wallace*, edited by Charles H. Smith (New York: Oxford University Press, 2008), 320–340, 337.

214. Charles H. Smith, "Alfred Russel Wallace, Past and Future," *Journal of Biogeography* 32 (2005): 1509–1515, 1513.

215. Ibid, 1511; and Smith, "Wallace's Unfinished Business," in *Natural Selection and Beyond*, 341–352, 346–347.

216. Ibid, 1512; and Smith, "Wallace's Unfinished Business," 350.

217. John van Wyhe, review of Michael Shermer, *In Darwin's Shadow*, in *Human Nature Review* 3 (2003): 166–168, 167.

218. Fred Hoyle, *The Intelligent Universe* (New York: Holt, Rinehart, and Winston, 1984), pp. 189–215, 248.

219. Hoyle, *Home is Where the Wind Blows: Chapters from a Cosmologist's Life* (Mill Valley, CA: University Science Books, 1994), p. 414.

220. Ibid., p. 421.

221. Nearly sixty years ago Duke University philosopher Glenn Negley wrote a savage attack upon the presumptions of cybernetics. "I am not sure what concept of *mind* it is that cybernetics is supposed to refute," he complained, "and even less clear as to what new or revolutionary theory of mind it is supposed to imply." See his, "Cybernetics and Theories of Mind," *The Journal of Philosophy* 48.19 (1951): 574–582.

222. Smith, "Wallace's Unfinished Business," 348.

223. Stephen C. Meyer, *Signature in the Cell: DNA and the Evidence for Intelligent Design* (New York: HarperCollins, 2009), pp. 86–92, 106–107.

224. Muriel Blaisdell, "Natural Theology and Nature's Disguises," *Journal of the History of Biology* 15.2 (1982): 163–189, 179.

225. Wallace, *The World of Life*, p. 416.

226. William Lane Craig, "Naturalism and Intelligent Design," in *Intelligent Design: William A. Dembski & Michael Ruse in Dialogue*, edited by Robert B. Stewart (Minneapolis, MN: Fortress Press, 2007), 58–71, 67.

227. See Alfred Russel Wallace, *Contributions to the Theory of Natural Selection: A Series of Essays* (New York: Macmillan, 1871), p. 372A.

228. Charles H. Smith, review of Martin Fichman, *An Elusive Victorian*, in *Journal of the History of the Behavioral Sciences* 31.1 (2007); pp. 97–98, 98.

229. Marchant, v. 2, 195.

230. C. S. Lewis, "The Laws of Nature," in *The Grand Miracle and Other Essays From God in the Dock* (New York: Ballantine Books, 1970), 51–54.

231. Norman F. Geisler, *Baker Encyclopedia of Christian Apologetics* (Grand Rapids, MI: Baker Books, 1999), p. 467.

232. Christopher Hughes and Robert Merihew Adams, "Miracles, Laws of Nature and Causation," *Proceedings of the Aristotelian Society. Supplementary Volumes* 66 (1992): 179–205, 207–224; and P. Hicks, "Miracles, Extra-Biblical," s.v. in *New Dictionary of Christian Apologetics*, edited by W. C. Campbell-Jack and Gavin McGrath (Downers Grove, MI: InterVarsity Press, 2006). The idea that miracles, in the truest sense, need not require a suspension of natural law is an old one. It was most thoroughly exposited by Nehemiah Grew (1641–1712) in his *Cosmologia sacra* (London: Printed by W. Rogers, 1701). More concerned with the purposes and results of "miraculous" events than their cause or causes, Grew specifically eschewed the invoking of

supernatural forces. For Grew, a miracle was "the extraordinary effect, of some unknown Power in Nature, limited by Divine Ordination and Authority, to its circumstances, for a suitable end" (p. 196). In this view, no law of nature is broken or suspended.

233. Geisler, p. 467.

234. Malcolm Jay Kottler, "Alfred Russel Wallace, the Origin of Man, and Spiritualism," *Isis* 65.2 (1974): 144–192. Kottler is not the only one to have argued in this vein. James Moore believes Wallace's theories can be largely understood from his conversion to spiritualism, and gives a supercilious dismissal of Wallace as the "spiritualist pretender to 'pure Darwinism'." See his *Post-Darwinian Controversies: A study of the Protestant struggles to come to terms with Darwin in Great Britain and America, 1870–1900* (New York: Cambridge University Press, 1979) pp. 184–190. Nonetheless, Kottler's remains the most thorough and balanced analysis of Wallace from this perspective.

235. Wallace, *My Life*, p. 120.

236. Ibid., p. 114.

237. Alfred Russel Wallace, "On the Habits of the Oran-utan of Borneo." *Annals and Magazine of Natural History*. 2nd series. 17.103 (1856): 26–32.

238. Slotten, 482–484.

239. Alfred Russel Wallace, *On Miracles and Spiritualism: Three Essays* (London: James Burns, 1875), 219–220.

240. Logie Barrow, "Socialism in Eternity: The Ideology of Plebian Spiritualists, 1853–1913," *History Workshop* 9 (1980): 37–69.

241. Ibid., 45.

242. Fichman, *An Elusive Victorian*, p. 286.

243. Ibid., p. 132.

244. Fichman, *An Elusive Victorian*, p. 317.

245. Wallace, *The World of Life*, p. 400.

246. See Appendix C. Also reprinted in *The New York Times*, October 8, 1911.

247. It is important to distinguish intelligent design from creation science. Intelligent design makes the comparatively modest though controversial claim that certain features of the natural world are best explained by an intelligent cause; creation science typically has prior religious commitments to supernatural agency in creation often framed within the context of the Genesis account. An example of the former is Meyer (see n. 223); an example of the latter is Fazale Rana, *The Cell's Design: How Chemistry Reveals the Creator's Artistry* (Grand Rapids, MI: Baker Books, 2008).

248. C. A. Patrides, "Renaissance Thought on the Celestial Hierarchy: The Decline of a Tradition," *Journal of the History of Ideas* 20.2 (1959): 155–166, 155.

249. Qtd. in Ibid., 163.

250. Fichman, *An Elusive Victorian*, p. 284.

251. Marchant, v. 2, 107–108.

252. Feisel Mohamed, "Renaissance Thought on the Celestial Hierarchy: The Decline of a Tradition?," *Journal of the History of Ideas* 65.4 (2004): 559–582.

253. William A. Dembski, *Intelligent Design: The Bridge Between Science & Theology* (Downers Grove, IL: InterVarsity Press, 1999), p. 45.

254. David Kohn, "Darwin's Ambiguity: The Secularization of Biological Meaning," *The British Journal for the History of Ideas* 22.2 (1989): 215–239, 238.

255. For more on the NOMA concept, see Stephen Jay Gould, *Rocks of Ages: Science and Religion in the Fullness of Life* (New York: Ballentine, 1999). Gould attempted to construct a kind of epistemological nosology that placed moral, ethical, and aesthetic concepts in one "Magisterium" and scientific, empirical data in another. Gould was attempting to address the old "science versus religion" debate by arguing that the two magisteria should not affect one another. It was precisely this walling-off of types of knowing into "upper" and "lower-story" thinking that Schaeffer found unwarranted and ultimately dangerous. For a thorough discussion, see Nancy Pearcey's *Total Truth* (Wheaton, IL: Crossway Books, 2005).

256. Francis A. Schaeffer, *Escape from Reason* (Downers Grove, IL: InterVarsity Press, 1968), p. 36.

257. Herbert Butterfield, *The Whig Interpretation of History* (1931; reprinted, New York: W. W. Norton, 1965), p. v.

258. John van Wyhe, "Alfred Russel Wallace: In a Court of His Own," *Evolution* 58.12 (December 2004): 2840–2841, 2840.

259. Ernst Mayr, "When Is Historiography Whiggish?," *Journal of the History of Ideas* 51.2 (April–June 1990): 301–309.

260. Steven Pinker, "The Cognitive Niche: Coevolution of Intelligence, Sociality, and Language," *PNAS 107*, suppl. 2 (May 11, 2010): 8993–8999, 8993.

261. Ibid., 8997,

262. Edward Feser, *Philosophy of Mind*, 2nd. ed. (Oxford: Oneworld, 2006), pp. 187–189.

263. Johan J. Bolhuis and Clive D. L. Wynne, "Can Evolution Explain How Minds Work?," *Nature* 458 (April 16, 2009): 832–833, 832.

264. Ibid.

265. David Berlinski, "On the Origins of the Mind," in *The Deniable Darwin*, edited by David Klinghoffer (Seattle: Discovery Institute Press, 2009), p. 435.

266. Feser, p. 78.

INDEX

Breinigsville, PA USA
18 January 2011
253560BV00001B/6/P